Terrace Gardening

Secrets and Techniques for Growing Herbs

(Simple Techniques to Grow a Sustainable Organic Garden)

Alethea Navarrete

Published By **Chris David**

Alethea Navarrete

All Rights Reserved

Terrace Gardening: Secrets and Techniques for Growing Herbs (Simple Techniques to Grow a Sustainable Organic Garden)

ISBN   978-1-998038-75-6

No part of this guidebook shall be reproduced in any form without permission in writing from the publisher except in the case of brief quotations embodied in critical articles or reviews.

Legal & Disclaimer

The information contained in this book is not designed to replace or take the place of any form of medicine or professional medical advice. The information in this book has been provided for educational & entertainment purposes only.

The information contained in this book has been compiled from sources deemed reliable, and it is accurate to the best of the Author's knowledge; however, the Author cannot guarantee its accuracy and validity and cannot be held liable for any errors or omissions. Changes are periodically made to this book. You must consult your doctor or get professional medical advice before using any of the suggested remedies, techniques, or information in this book.

Table Of Contents

Chapter 1: Creating A Mindset For Terrace gardening

"Perfectionism is the unwillingness to be inclined."

Around 604 BC in Babylonia, the "putting gardens" is probably decided. These gardens have been now not planted in putting baskets however alternatively have been planted on top of stone columns. The flora may want to hold over the columns and in doing so might likely resemble "setting gardens."

In Greece, round 375 A.D., rooftop gardens can be placed. These rooftop gardens consisted of brick packing containers which have been planted with an series of vegetation. These containers had been then positioned on the roof wherein no longer some thing else might increase.

Today, the interest in Terrace gardening has grown thru leaps and limits. Urban development, adulteration in food, loss of farming, and these days developing spaces & corners in your own home as tested on Instagram and Pinterest are the few reasons which have spurred this recognition.

Terrace gardens have come an extended manner and traveled the globe. Today, Terrace gardens can be placed on the floor, setting above your head, to your balcony, and your rooftop. So jump each toes into this vintage planting craze and create your definition of Eden via Terrace gardening.

Why a attitude is needed?

With the introduction of social media, every idea that become relying on imagination a few a few years inside the beyond is now with out issue available or maybe executable.

Terrace gardening is likewise slowly turning into like a style. There isn't something incorrect with that however in case you want to make a Terrace lawn (repeat... a lawn) now not only some flora lying here and there, you should set your expectancies in advance than you start or try and amplify your Terrace gardening adventures. This might in all likelihood sound like me pronouncing, "Terrace gardening is not clean" so stay with me here and look at on.

10 exceptional thrilling motives to increase a Terrace lawn

1.Versatile: Terrace gardens will will let you growth flora on a balcony, terrace, interior on windowsills, or rooms with ok mild, bathrooms, railings, pergolas, and plenty of others.

2.Variety: You can increase plants that won't be appropriate to increase on your lawn soil and expand flowers subsequent to each unique regardless of the reality that they will have precise soil needs.

3.Mobility: Plants in pots can with out issues be moved to suit your goals or to a extra suitable sunny or shady area inside the route of the day. If you convert your private home, your plants go with you in boxes.

4.Looks: Organize flora with out trouble to suit your own home decor

5.Flexibility: Rearrange flowers to match the season or your flavor.

6.Fewer Disease & Pest Problems: Of direction muss lesser as compared to a land garden or a farm concern.

7.Reachable: Convenience of simplest attaining out or taking a few steps to glowing homegrown herbs, stop end result, and vegetables.

8.Scalable: As you research and benefit enjoy you may expand your Terrace lawn.

9.Recycle & Repurpose: Terrace gardening promotes a whole lot of recycling and repurposing of things at domestic in any other case discarded to harm our surroundings.

10. Positivity: Having a Terrace garden lets in you keep calm, reduces thoughts chatter, and therefore you're extra present in each second.

Are the ones motives perfect enough to inspire you!!

But is it smooth?

There isn't any easy way to this so permit me will let you realize some topics in advance than you answer the question above.

Let's do a easy workout first. Close your eyes and visualize your house, then pass a piece father and your neighborhood, then trees round your locality, then nearest park, then nearest farmland, then nearest plant life, then nearest hill, then a forest, and agree with your self taking walks in this wooded area wherein you can pay attention the sound of leaves on the ground as you are on foot, feeling branches and leaves touching your body, the solar playing cowl and are attempting to find the various timber. One final thing, in this visualization, now attempt to find out a container with a plant in it.

Of course, you did now not discover a location with plant in it inside the forest. That's their herbal habitat, the floor, the trees, the fallen leaves, animals and their droppings, birds ingesting cease result and bees over plant life, fallen timber, lifeless animals & vegetation getting evidently decomposed, a water body like a lake, fall or river close by or rains.

A plant in a woodland by no means has to fear about its growth, watering or over watering, pruning, pollination, fertilization or overfed, colour or solar, roots or root-positive. Because there may be a beautifully balanced surroundings that looks after the whole lot that is favored for the plant to stay on and extend.

On the alternative hand, whilst we're developing a plant in a discipline, your plant is one hundred% depending on you. A plant will continuously be your little one all of the time.

You can in no way incorporate nature in a box, therefore you should do your great to replicate or provide comparable conditions. Sadly that is in which mistakes display up and those surrender. My device for terrace gardening is constructed on the inspiration of 100s of errors that I made at the same time as doing Terrace gardening and commonly I felt that it's far really tough however best one difficulty that helped me persevere and therefore I succeeded in terrace gardening.

So going again to the query. Yes, it's easy if you have a proper system in vicinity. Otherwise, there may be an entire lot of experimentation, errors, and disappointments that I honestly have skilled and I do not need you to replicate that.

Chapter 2: Secret Ingredient For The Success

"Every unmarried time you professional some issue proper to your lifestyles, you loved and harnessed love's brilliant strain. And every single time you enjoy some aspect now not correct, you didn't love and the stop give up result changed into negativity"

Trust me after I say this irrespective of what you do to flourish your terrace lawn if this one detail is missing, you could now not be able to be successful. Let me let you know why I suppose that mystery element works.

Have you ever questioned why a few humans can keep 100s of discipline-grown plant life on the equal time as others battle to control best just a few? In the fraternity of gardeners, such humans are typically referred green thumb, brown thumb, black thumb. And to apprehend all three you just need to understand the only Green Thumb

According to Merriam-webster dictionary, a inexperienced thumb is someone having an uncommon capacity to make flowers develop. Now you could wager what brown and black thumbs are.

But, I have my definition of a green thumb. Stay with me right here.

Plants are people

Plants are greater than simply mute dwelling beings.

Summer Rayne Oakes, an metropolis houseplant expert, and environmental scientist is the icon of fitness-mindcd millennials who want to carry nature interior, consistent with a New York Times profile. Summer has controlled to broaden 1,000 houseplants in her Brooklyn apartment (and they will be thriving!)

A 1000 plants in an apartment is virtually too top to be actual and also you need to be questioning that she is a green thumb. She ought to be performing some aspect so right that she has an entire jungle in her condominium in the center of an city metropolis.

What is her thriller?

Well, she strategies her relationships with vegetation as deliberately as despite the fact that they were humans.

So, what is the simplest factor of life that makes people increase and flourish? It is the equal components for plants.

Plants love tune

There have grow to be a take a look at executed on April 22, 1981, at Regional Research Center Jhansi (UP) approximately the Effect of mantras on people & plant life[2] . Here's the precis from the have a observe

"The conventional texts of Indian foundation record the impact of Mantras on plant life and animals. Ayurveda additionally acknowledges the significance of this realm of drugs. The creator inside the direction of his severa experiments on plants determined that these from the diploma of seedling to maturity are laid low with positive types of sound waves, in particular the Mantras. This test well-knownshows that the vegetation have proven a awesome reaction to this form of unique sound waves concerning increase. Their efficacy in curing the illnesses and lots of others"

Another have a have a look at changed into posted in Sept 2019 within the International Research Journal of Engineering and Technology (IRJET) about the EFFECT OF VEDIC CHANTING ON PLANT GROWTH PARAMETERS [3]. Here's the summary from the have a test

ABSTRACT: Music treatment is one of the high-quality and most popular remedy in modern years, it had a superb, high-quality have an impact on on our physical and physiological situations. The vibrations and frequencies crafted from the track can capable of be felt with the resource of manner of vegetation and reply regular with that. They develop at a few unique frequencies and at a few special tremendous frequencies their growth is stunted. It is primary that the plants which might be exposed to Vedic mantra chanting have a incredible impact on growth, leaf duration, and inter-node and buds. Two gadgets were determined on on the moong bean plant – one handled with Vedic mantra and the second set untreated i.E. Manage. Amongst the two sets, mantra-dealt with flora verified higher growth parameters like germinated amount of seeds, the peak of the plant, range of leaves, root duration, chlorophyll content material cloth cloth, it is more for the chant-handled plants and decided have been an lousy lot less for the manage set.

When you suspect of music, there are mind of pride and happiness. A plant is glad with the tune and so are plant humans. The mystery detail is an emotion each person connect with.

the secret detail is...

. It could not be an overstatement to mention that vegetation have emotional responses. So going once more to our query "What is that one detail of life that makes human beings broaden and flourish?"

LOVE: It is love that makes humans expand and flora are humans.

the new definition of a inexperienced thumb

A inexperienced thumb is a person who loves plant life like people.

But does each person love flowers? Not real...most people love the idea of proudly owning a plant in its full glory, be it flowering, fruiting, or creating a statement for the space it's miles positioned in. In my and tens of thousands and heaps of gardeners' experience, at the same time as you certainly love vegetation, you need them like babies like youngsters.

When you've got a toddler, you try and provide the first-class surroundings to develop, feed, care for them. When they are ill, you display them intently every unmarried day to ensure their recovery.

The metaphor that vegetation are infants all of the time is so real due to the fact like infants they can not let you recognize what they're feeling, whether or not or not or no longer they are unwell or healthful however they'll get hold of this exquisite ability to provide you with signs and symptoms and indicators for any motion deemed important.

That's love and just like infants in pass lower back, you get love & the purest form of pleasure.

You need to experience need to harness its strength. With this energy, you are unstoppable in a few component you want to do.

Love is the riding pressure of nature.

Chapter 3: What All You Need

"Gardens aren't made through way of approach of making a tune 'Oh, how beautiful,' and sitting in the colour."

Papa!! "Why there are pits in our pot's soil?" I requested.

He said, "It need to be a mouse or cat gambling spherical."

When I became a toddler, we used a plastic mug to water my vegetation. Watering this manner induced the ones small pits in the pots and that's at the same time as located out that I want some thing to gently water my flora with out making those pits.

In this monetary damage, you'll discover about matters that you need to start with and hold a terrace lawn. Let's begin with the plain one.

SPACE

The maximum apparent one, you want place. For sake of understanding I certainly have divided region into the subsequent commands:

1.Large regions like a terrace or a large balcony, rooftops, and so on.

2.Medium areas like a medium balcony or a nook or a part of a big balcony.

three.Small areas like small balconies in immoderate-upward push houses.

4.Really small regions like a French window sill

five.Ultra-small areas like a small window

6.Hanging areas like projection roofs over home domestic home windows, balconies, and so on.

A small balcony or corner in a terrace is a good area, to begin with, your Terrace gardening which can increase over the years.

trends of tremendous space for Terrace gardening

1.Water drainage: It's a no brainer but I can not emphasize masses how crucial it's far. Make high high-quality your space has water drainage simply so there may be no stagnation of water. This is quite vital :

1.To save you roof seepage.

2.To prevent mosquitos from laying eggs.

three.Accumulation of soil worn-out from containers over time can also damage the floor/tile texture.

four.May purpose conflict along aspect your neighbor under your apartment if it's a balcony or window.

2.Waterproofing: If you are planning to have a Terrace lawn so that it will growth constantly (and it's going to even as you pour your love in it) over time. Please ensure there is a few kind of waterproofing on your terrace/roof. I apprehend the manner it looks like at the equal time as your newly designed fake ceiling starts offevolved offevolved leaking water inside the middle of your mattress room☺.

three.Well lit: Space have to be well lit with direct or indirect sunlight. This can even impact your choice of plants forTerrace gardening see Chapter 10. Secret To Choosing The Right Plants.

four.Provision for Shade: East-going via area gets preserve of morning mild whilst West going through region will achieve afternoon mild that is hotter and brighter. When critical you want to be able to cowl the space with a shade like a green lawn net.

5.Ease of cleansing: You may need periodic cleansing rituals to hold the gap tidy. Though you need to maintain this in mind on the same time as packing this location with containers.

6.Safe for youngsters/pets: If you have got got kids or pets, please make sure that every time required you could make your vicinity inaccessible for your children or pets. This can be finished with teenager's/pets door, doors with a grill, or only a door. It's very crucial for his or her safety and moreover the protection of your plant infants ☺ .

One of my terraces full of flowering and ornamental plants

One of my terraces

Besides the plain, allow me will will let you realize a few things that you could want to begin your terrace lawn. You don't must non-public they all right away therefore I sincerely have classified them as follows:

1.Beginner

2.Intermediate.

3.Advance.

Think of novice, intermediate, and extend as stages of the way your terrace garden grows. It isn't your enjoy that I mean.

Note: Feel free to shop for these items every time you want them.

newbie

Assuming that you are absolutely beginning your Terrace lawn. Here's what you can need.

Containers

It's a no brainer, you want containers to offer your plants a domestic.

Covered in high-quality depth in Chapter 4. Secrets To Choosing The Right Container

Potting Mix

To increase your plant you need a medium for them to amplify.

Covered in super intensity in Chapter five. Is It Potting Mix Or Soil? What's The Secret?

Watering can

A 3lit or 5lit water can with a detachable shower head is a must-need to nicely water your plants. Watering ought to make watering clean, effective, mess-free, and amusing.

Face masks or cover

For the ones occasional duties even as you is probably uncovered to dirt you may see Chapter 14. Gardening, Self care & Safety for why it's so critical.

Large plastic sheet

My family loves this concept. Working with potting combination, transplanting, repotting is a messy project, and cleaning the space in some time may be very aggravating and tiring. Using a large and accurate-excellent plastic sheet to cover the space and artwork on top of it is extremely beneficial in maintaining your space clean and tidy.

Planter stands or a jugad

Whenever viable I strive no longer to region my containers without delay on the floor. You should purchase or custom-made some simple stands however in case you need to wait till you have received greater experience then truely area damaged tiles/bricks to raise the container degree from the floor.

Warning: Whenever you buy/custom-made planter stands of iron, ensure that its legs are typically made the use of cast-iron like an perspective bar, solid bar, or a colourful bar and no longer the pipe. And in no way offer rubber (rubber shoes to prevent scratches) assist at the bottom. The water will gather in the ones shoes or pipes and it's going to ultimately motive rust and rot.

A awful planter stand

Some gadget from the kitchen

A kitchen is a tremendous vicinity for garden substances and you could effortlessly get the subsequent:

1.Spoons

2.Fork (useful for tilling, root pruning, root clearing)

three.Scissors

4.Sieve or stainer

Intermediate

Once you could manipulate your initial Terrace gardening setup, I propose you to have the subsequent gadgets on your possession.

1-liter spray bottle

A real excellent spray bottle is beneficial for spraying pesticides and fertilizers. A 1-liter duration is handy to begin with.

Long hose pipe

Invest in an awesome-splendid hose pipe that is weatherproof, does no longer bend or twist. Very beneficial and saves an entire lot of energy watering the plants.

Long lawn shower head

Watering is that this type of pleased interest and having a fountain-style shower head not best makes it greater amusing however furthermore helps in cleaner and uniform distribution of water in a discipline. It does now not disturb the soil in contrast to using a mug or direct hose pipe. Your plant toddler will love this shower.

Small Trowel

While you may be proud of having your hands dirty but having a accessible trowel make it very green to art work with potting mixes.

High-tremendous secateurs/Pruning knife

What a extremely good knife is for a chef having a top notch secateurs/pruning knife/pruning shear is for a gardener.

Neem Oil

Buy neem oil for your lawn. It is my lone soldier for pest manipulate and management. Buy a great oil that's unadulterated with chemicals or great oils.

Seaweed extract

In easy phrases, that is a tonic to your plants. In clinical phrases, it's an extraordinary source of micro-vitamins (realise greater approximately plant nutrients in Chapter 6.

Secrets Of Feeding Your Plants) to your vegetation and consequently it's an extraordinary readymade natural increase promotor to your Terrace garden

Advance

In my humble opinion, there can be in no way a complicated diploma of gardening if you keep analyzing and increasing your garden, you could typically pass on to a superior degree however when you have the following equipment, it way you've got already got a top notch lawn to control

Gardening internet with the manual form

Plants love moderate but we need to guard them for the duration of warmer months and that's why we want garden nets, see Chapter nine. Secret To Fantastic Care to apprehend more about Gardening Nets

five-liter spray bottle

Unless you want to keep in thoughts spraying your lawn a cardiac hobby for weight reduction ;-), it is better to have a larger spray bottle. It saves pretty a while and electricity.

Vermicompost

Made using trojan horse castings. An superb herbal issue for potting mix and also awesome as a fertilizer additive.

See Chapter 5. Is It Potting Mix Or Soil? What's The Secret?

Neem Cake Powder

A dried byproduct of oil extraction gadget from neem give up result. An first-rate natural additive for potting mixture and moreover extraordinary for pest control.

See Chapter five. Is It Potting Mix Or Soil? What's The Secret?

Cow Manure

Decomposed cow dung. An superb herbal detail for potting aggregate and additionally terrific as a fertilizer additive.

See Chapter 5. Is It Potting Mix Or Soil? What's The Secret?

Mustard cake powder

A dried byproduct of oil extraction manner from mustard seeds. An terrific herbal additive for potting combination and moreover terrific as a fertilizer additive.

See Chapter 5. Is It Potting Mix Or Soil? What's The Secret?

Bone meal (sterilized)

Slaughter house waste. An extremely good herbal additive for acidic potting aggregate and moreover amazing as a fertilizer additive.

See Chapter five. Is It Potting Mix Or Soil? What's The Secret?

Humic Acid Liquid (Optional)

An with out issue to be had organic liquid fertilizer that helps in improving the exceptional of soil, it enables in stronger roots of the seedling. I use humic acid in maximum of my plant life once I'm repotting or transplanting.

Multi-sample water spray gun

This spray gun is my all-time preferred. It lets in to water the plant, smooth leaves & dust. Since the drift of water may be stopped with the lever it additionally permits saves plenty of water. I strongly propose you buy this tool.

There are not any suggestions to purchasing for any of the above gadgets so revel in loose to buy them as you deem crucial.

Lets us stay deeper into choosing a home for your flora i.E. The field.

Chapter 4: To Choosing The Right Container

"Every container for your lawn is a garden in itself."

Every vicinity in your Terrace lawn is a garden in itself. Each subject has its surroundings. Pause for a 2nd and bear in mind what I really said and you'll be surprised how authentic it's miles. And this is why it's miles of vital significance that you choose out the proper discipline.

In this financial disaster, I will share all of the secrets and techniques and techniques and strategies of bins that I truely have located over a few years and the manner you could choose the top notch ones. But experience loose to make your private choice and test.

Avoid slender necks

Any place that has a stomach like earthen pot (matki) and a slim neck isn't always so nicely preference for a container due to the subsequent reasons:

1.Difficult to take out the plant for repotting without damage to the inspiration ball.

2.The narrow neck can also preclude the airflow to the soil which also can lead to root rot or bad increase.

three.If you can not take out the plant, you could have to break the pot. While you can discover it good enough to interrupt a few earthen pots in a yr or so however it is probably a waste if the ones pots are expensive ceramics or similar cloth.

four.Hoeing for potting aggregate topping is tough.

But it seems so super

It is a private choice but advantageous sorts of such round containers do add to the aesthetics of the gap (you can have seen those brass bins in resort lobbies). So in case you insist on using it then do as follows:

1.Make advantageous your potting mixture is nicely-tired and loose textured See Chapter five. Is It Potting Mix Or Soil? What's The Secret?.

2.Do now not immediately plant it alternatively plant in a normal container a piece smaller than the spherical pot and positioned the pot indoors it.

three.For cactus and succulents, this kind of pot is o.K. Because the potting blend is extremely good well-worn-out and free textured

Plastic vs clay

Sooner or later you can simply discover yourself stuck in the seize 22 scenario or everlasting debate among plastic vs clay pots. Why is that?

Disadvantages of clay pots

1.Breaks without problems, in particular if moved spherical plenty. It's hard to repot vegetation with out breaking the planter.

2.Drainage issues, attempt drilling holes in a clay pot without breaking it.

three.Plants require extra frequent watering.

4.Clay is vulnerable to cracks

5.Broken clay can produce sharp edges that aren't steady round kids or the general public.

Disadvantages of plastic pots

1.Attracts and keep warmth

2.Cheap plastics can be very brittle and skinny, without problem chips and cracks

3.Cheap plastic is susceptible to fading in daytime

four.Cheap plastic planters don't remaining greater than multiple seasons

Which one to pick out

Clay pots were there considering the begin of field gardening however they've preservation problems consequently I might recommend going for plastic/resin-based bins, and masses of others. To overcome their dangers invest in correct incredible non-fancy planters (Don't you observed the complete fancy factor is your plant ☺). There are also UV-dealt with planters that last lengthy inside the open and replicate warmth.

Besides that, you get a whole lot of range in non-clay planters as hangings, railing, and vertical planters, and lots of others.

No flat backside

Do not buy pots/planters with a flat bottom. A flat bottom planter has no floor clearance from the lowest of the planter or no legs to raise the planter from the floor. A flat backside planter is not an brilliant preference due to the truth

1.It obstructs airflow via the planter and does not allow the gravity art work for water drainage.

2.Overgrown roots may also pinnacle outside the drainage holes due to the fact there is moisture retention at the lowest. This may moreover bring about horrific growth, root rot, contamination, and so on.

three.It makes it tough to easy the space. Even worst if your water supply has tough water it can damage your floor/tile texture.

Flat backside vs ground clearance

If you have got already were given one

If you have got already got flat-bottom planters you may enhance them above the floor thru using:

1.Putting them on some planter stand.

2.Placing them on bricks/quantities of stone slabs

Getting cheaper bins

Every gardener spends plenty of cash purchasing for suitable commercially available pots/containers. There are numerous blessings to it:

1.Easy to preserve consistency within the region.

2.Ready to apply, no drilling holes, and so forth. For drainage.

3.Lots of range, style, colour, duration shape.

four.Usually have all company necessities like ground clearance, markings for potting mixture degree, drainage holes, UV treatment, and so forth.

While all of the above benefits aren't any-nonsense it has the capacity to make a big dent on your pocket in case your lawn continues to grow specially big-size boxes. Therefore it's important to have less high-priced but right pleasant options, underneath are my private choices:

Paint buckets

Paint buckets are my personal favored because of the fact I get them almost free of charge.

What makes paint bucket a wonderful choice for Terrace gardening:

1.They are made the use of superb extremely good plastic to face up to the weight at some point of transit and motion.

2.They are available an entire lot of sizes.

three.The larger ones additionally have handles. A little tweaking and also you join ropes as handles on your large plant infants.

4.You can get them free of price (nicely almost) if your friend's or relative's home goes through a paint procedure all you need to do is in a polite manner ask or convince them that it is of no need for them ..LOL!

five.Diwali is the top notch time to hunt for paint buckets.

6.For unique instances of the 365 days, you could call your paintwala to installation a few for you at an inexpensive fee.

7.Every town has a scrap market in which you may find out paint buckets. I sold a 20-liter bucket for 80Rs each, a similar duration business pot may have rate no much less than 200Rs

The handiest hassle with those buckets is that you want to drill the drainage holes manually and moreover provide something for ground clearance.

Blue barrels/drums

My 2d preferred is the ones large blue barrels that are used to deliver and shop severa liquids for enterprise uses. They have all of the benefits of a paint bucket plus they come in medium to massive sizes making them suitable for planting timber. They moreover include handles which makes them less complicated to transport round.

NOTE: These barrels save you from repotting for many years otherwise it'd had been slightly difficult to carry out that annually due to their round bellies.

Same as buckets you want to drill drainage holes.

Old water garage tanks

If you have got vintage water storage tanks , they make superb packing containers to be used as gardening beds. You can also want expert help to interrupt up them in 1/2 of.

Large length drainage holes

If you buy a commercially to be had field, there are 3 opportunities:

1.It may additionally moreover have drainage holes.

2.It may not have drainage holes but have provisions to make.

3.It can also have drainage holes but both they will be small or no longer sufficiently sized.

If you are fortunate you could get an incredible splendid place that has accurate period and range of drainage holes. Drainage holes are the most essential function of your area. The fitness of your plant in long term in huge part is based upon on.

Whenever I actually have to shop for a readymade planter I make certain that it has an tremendous substantial style of huge drainage holes. If the size is simply too small I increase it. For paint buckets and barrels, I create the ones massive holes myself.

Why big period drainage holes are higher

1.Smaller holes can also get clogged over time with potting mixture (of some aspect kind).

2.Smaller holes do no longer artwork properly with gravity to drain the water out.

3.Smaller holes prevent the airflow which again does not permit water to come out.

4.There is not any manner to de-clog smaller holes besides to repot the plant.

five.Large length holes triumph over all the above stressful conditions.

6.You don't need gravel or stones at the bottom with the large hollow approach that I will percent below.

Diy best drainage with big holes

Follow the steps to create exquisite drainage with large size holes:

Tool Required:

1.Drill device

2.Hole observed attachments

three.Plastic mosquito internet

4.Safe tool like goggles, masks, gloves, and lots of others.

5.A black cool animated film pen

6.A scissor

Steps:

1.You may need a drill tool and a hollow observed. If you do now not have, then you want to discover a person who can try this for you.

2.Remember to put in your safety goggles or at the least ordinary goggles, mask, gloves, and so forth. In case you are doing it with the resource of yourself.

three.Place your subject the other way up.

4.Use a 1-inch hole noticed and drill a hollow within the middle of the lowest. Note: The length & quantity of holes is based totally upon on how big the field is.

five.Then use ½ inch hollow observed and create extra holes. The simplest problem to be careful of is that the massive holes need to no longer be too near otherwise the bottom will break with the burden.

6.Once you're down with all of the holes, unfold the mosquito internet and place the container on it. Using a groovy animated movie pen draw a circle on a mosquito net with the sector's bottom.

7.Cut the circle and area it inside the subject's bottom. Fix the rims and gaps if any by the use of pressing them firmly.

eight.Your subject's drainage is prepared ☺

Now you may be asking "Wouldn't such type of drainage may additionally moreover additionally bring about lack of nutrients from the potting combination?" Yes, it will but not without a doubt this type any well-worn-out potting mix will leach its vitamins with drainage over the years that's why we use additives and fertilizers. See Chapter 6. Secrets Of Feeding Your Plants.

I pretty suggest which you discover ways to use a drill device on diverse surfaces, it's far the most critical DIY device that you could examine.

No gravel, pebbles, or rocks at the bottom

You might likely understand that traditionally many gardeners have been putting gravel or rocks at the bottom of the sector for precise drainage, a few no matter the truth that do. Rocks don't enhance drainage; instead, they decorate the perched water desk[4] in the direction of your plant's roots. The vegetation emerge as sitting in too wet soil, growing the opportunities of plant root rots and also you've wasted valuable pot space with gravel that's doing no right.

It's better to popularity on selecting the proper potting mixture that is porous sufficient.

Placing a bit of best plastic mesh over the planter's drainage hole will assist maintain soil from washing out and received't save you water drainage.

Size, form & shade of boxes

Social media is showcasing a plethora of planters/boxes of numerous styles and sizes. You will discover images of balconies, patios, terraces so cute that it's far impossible to stand as much as. In my enjoy, not all the ones planters/boxes are practical and/or super on your plant's fitness. Besides that, your plant may not whinge approximately the planter it is growing in but it can now not healthy the complete association.

Size

The duration of your container relies upon at the type of plant you're approximately to broaden but proper right here are a few pointers to choose the right duration:

1.Never pick out out a period based totally on the preliminary u.S.A. Of the plant that you purchase from the nursery. Choose a bit larger. Check with the nursery guy about the increase nature of the plant. For example, pothos also can develop fast and outgrow eight-inch boxes at the same time as a ZZ plant would possibly do certainly exceptional for a 12 months or 2.

2.Size and shape do no longer generally mixture nicely: Sometimes the form of the planter is so attractive that it's smooth to overlook the size requirement. Don't try this.

three.Don't buy an oversize area hoping that subsequently, your plant may be huge in length. Always pick out a medium to a large however not large planter. Though I may additionally want to recommend a big field in case you are growing wooden or fruit timber.

Shape

Plants don't care approximately the shape of the box till they battle to extend. Below are my pointers while choosing nonstandard shapes:

1.Stability: No recall what the shape is constantly make certain that your planter is stable at the same time as full of potting mixture and planted. This is in particular actual for railing planters, setting planters as they may be hazardous inside the occasion that they fell.

2.Material Quality: Be careful of the cheap great material used for fancy planters as it can increase cracks and can be broken if fall.

3.Hanging mechanism: Hanging planters of each type depend on the weight-bearing ability in their putting mechanism. Hanging baskets with plastic setting hooks receives twisted & fall inside the occasion that they have got extended exposure of solar in excessive temperature every day. I generally choose out a mild-weight potting blend. You can also without troubles create strong putting assist the use of GI (Galvanized Iron) wires. M

DIY twine placing mechanism for strong resource

Color

Colorful bins is an difficulty of personal desire however I suppose I can though let you know a detail or two approximately colored bins:

1.Avoid wonderful or neon-themed shades in case your flowers are despite the fact that younger and however to have to turn out to be busy & blooming. Otherwise, all you could see are the ones sun sunglasses now not the herbal splendor of your vegetation.

2.If your plant is a lovely flowering plant or has foliage with shades of more than one colours and textures, then please use primary coloured planters like terracotta color, black or white, or gray. This will spotlight the plant and its splendor.

3.Use colorations close to nature. Go discern ☺

4.If I add colorations to my lawn except flowering flowers and foliage I do that with the aid of such as gardening add-ons and depart my planters to simple colorations.

Hiding the ugliness

Sometimes the container you chose is top notch suited in your vegetation however they may no longer be the maximum eye-attractive detail to your Terrace lawn. Here are a few easy and smooth techniques to offer cover to such planters:

1.Surround it with different smaller bins to cover it.

2.Paint them in some extremely good natural sunglasses.

3.Draw on them. You also can draw some smooth styles to make it appearance appealing.

four.Make some planter cowl the use of pallet wooden (there are dozens of tutorials available on-line.)

5.Use a planter cowl basket (you may find out it in-shop like Westside, home center, and so on.)

6.Glue a cotton rope across the planter to add a rustic look for your garden.

Grow baggage: An developing fashion

Well because of the fact the call indicates, it's a bag. A bag you could fill with potting combination and grow vegetation in them. As increasingly human beings are trying to expand (specially veggies) on their terraces/balconies, increase bags provide pretty a less highly-priced opportunity to conventional packing containers.

Why did they emerge as famous?

The well-known opinion about expand baggage is "Grow baggage are lighter than pots and so they may be simpler to transport, maintain, and shop. They are pretty cheap compared to ordinary boxes, and they're to be had in a number of shapes and materials. You can also DIY grow bags".

Types of broaden luggage

Grow bags were in the starting crafted from HDPE/Plastic but these days they're moreover to be had in jute or material (fiber). The latter is a chunk expensive in evaluation to HDPE bags.

An HDPE broaden bag

Pros of increase bags

1.Wallet-excellent: No doubt this is one of the maximum crucial motives, they have got emerge as so well-known.

2.Ease of use: It's probably a incredible brief way to installation a terrace lawn with out lots trouble.

three.Better Soil Aeration & Drainage: Grow baggage are breathable because of this higher roots and ordinary top development of the plant. This however is exceptional viable the usage of Fabric/Fiber bags.

Cons of develop luggage

1.Durability: Total is predicated upon at the incredible of the fabric and environmental situations. Fiber baggage also can very last longer than HDPE.

2.Excess water intake (Fiber bags): They want more watering than ordinary packing containers because of aeration. Miss watering on warmer days and your plant may be at danger.

3.Stability: Larger increase luggage, specially rectangular & oval ones don't preserve their shape at the same time as they will be entire of potting combination. They need more help systems like a body.

4.Mobility: Grow luggage can emerge as as heavy as plastic packing containers if a conventional soil-primarily based completely truely potting mixture like lawn soil+sand is used. Also, huge bags are susceptible to tear in material due to inclined handles. So they will be no longer as smooth to move as it appears.

5.Ground clearance: Since there is no floor clearance, you would want a stand or drainage cells to area your luggage. This in fact shoots up the general price range

6.Not Eco-Friendly: HDPE luggage aren't inexperienced and no longer all kinds are biodegradable or recyclable. Some fabric luggage are made the usage of fiber derived from plastic water bottles, the ones are properly nicely worth checking.

Turmeric growing in Fabric grow bags crafted from recycled doggy bottles

Even with greater cons than experts, I would possibly despite the fact that say grow luggage are nicely for terrace gardens however won't be for each terrace lawn.

Choose them correctly!

Chapter 5: Is It Potting Mix Or Soil? What's The Secret?

"We can also count on that we are nurturing our garden, but of route it's our lawn that is nurturing us"

When you communicate to the soil in gardening additionally it is a confusion among a potting mixture and gardening soil. So what's a potting combination and why it isn't soil and it need to not be known as so?

A little bit of soil technology

Plants broaden flawlessly nicely in healthful, nicely-amended garden soil so why can't we use the equal soil in our bins?

One purpose is drainage, at the ground in a garden bed gravity acts at the soil from deep underground, allowing extra water to empty down and far from plants' roots. But in a field, the strain of gravity can simplest pull water to the lowest of the field, wherein it can or may not discover a drain hole to break out.

At the identical time, water is likewise drawn upward and is held inside the soil by way of capillary movement – the equal strain that attracts water slightly up the straw because it sits in a cup of water. (Remember floor clearance feature of a superb discipline?) And the thinner the straw, the better the water is pulled. Similarly, water gets pulled upward into the pores of soil – specially in the tiny pores of our clay-heavy soils – wherein it's held. So even with an incredible drain hole, a number of the water absolutely obtained't have the capability to interrupt out in a container full of lawn soil. And an excessive amount of water within the soil for too prolonged can lessen off air drift across the roots, that could pressure or possibly kill a plant.

Another reason is contamination of soil through the usage of chemicals, sewage, chemical fertilizers, and pesticides. To apprehend why a potting blend over garden soil is maximum appropriate, understand what a potting mixture is.

Potting blend

Like soil, potting combination gives help & vitamins for plants however it isn't soil. It's a increase medium formulated specifically to assist plant life stay happily in pots forming a snug surroundings to guide their boom.

Instead of lawn soil, agency potting mixes include a combination of peat moss or coco peat, composted cloth like vermicompost, tree bark, cow manure, perlite, and vermiculite – and once in a while a few one-of-a-kind components, counting on the combination.

Because of those materials, a outstanding potting combination holds truly the right quantity of moisture while letting the more drain away with out problems. It also continues an ethereal, "fluffy" form whether moist or dry. Potting combo simply makes box lifestyles plenty less complex for flowers.

Soil plenty much less potting combination

In layman's phrases, soil-plenty lots less potting combination (media) is a few aspect that doesn't have dust/garden soil/mitti as an aspect.

Advantages of soil-a good deal plenty less potting blend over the soil-primarily based absolutely

1.Planters with mixes weigh 50-70% lighter than traditional mixes.

2.Plants broaden higher due to the fact its fluffy and loose nature.

3.Roots develop higher and faster as they have got more breathing space.

four.Potting blend encourages moisture retention and further water drainage, roots of the plant breath higher.

five.Adding periodic manuring or natural additives like bone meal, mustard cake, neem cake, and so on. Is trustworthy and inexperienced. You no longer regularly want hoeing. This can mimic the ground of the soil inside the wooded location hence help assemble a outstanding environment for the plant to increase.

6.Goes without announcing that this mixture holds nutrients longer than regular soil-based blended.

7.To a few extent, you may make certain that your mixture will have lesser fungal, pest, or insect infestation as ordinary garden soil/dust can be infected.

8.Re-potting will become less complicated and green.

nine.Removal of weed is an awful lot less hard.

Ingredients of potting combination

There aren't any policies with regards to developing a potting blend but sure herbal factors had been in use for any such long time in gardening for their apparent advantages. Let's apprehend why I decided on the ones herbal materials for my potting blend recipe

Vermicompost

Vermicompost (vermicompost, vermiculture) is the manufactured from the decomposition way the use of numerous species of worms, commonly purple wigglers, white worms, and specific earthworms, to create a aggregate of decomposing vegetable or food waste, bedding materials, and vermicast (bug poop).

Vermicompost is a extraordinary deliver of natural carbon and distinct important nutrients for vegetation.

You can with out hassle purchase vermicompost in recent times from your network lawn maintain or nursery.

Nutrition Composition of Vermicompost [5]

You can also additionally use garden compost or home made compost as well. Homemade compost is a first rate manner to lessen kitchen waste.

Cow manure

For many years cow manure (gobar khad in Hindi) has been used considerably in Indian farming strategies and gardening due to its ease of availability.

Cow manure is made of digested grass and grain. Cow dung is high in herbal materials and rich in vitamins. It consists of approximately 3% nitrogen, 2% phosphorus, and 1% potassium (three-2-1 NPK).

Cow manure also carries immoderate stages of ammonia and probable risky pathogens. Therefore it need to be aged or composted in advance than its use as cow manure fertilizer.

Cow manure is an financial and effects to be had source of herbal carbon, organic depend, and extraordinary plant vitamins. Which makes it an exceptional difficulty for my potting aggregate.

Coco peat

Coco peat, also known as coir pith. Coir fiber pith, coir dirt, or definitely coir, is made from coconut husks, which is probably byproducts of different industries that use coconuts.

Coco peat does not have any nutritional fee but it's far completely natural and has notable water protecting functionality. It also acts as a neutralizer for heavy mediums like vermicomposting and cow manures.

Coco peat is used to feature in addition lightness, water retaining capability, and conditioning to the potting combination.

Commercially available cocopeat blocks have a higher degree of salt carries which may moreover have an effect at the tremendous of your potting combination and consequently eventually ward off the growth of your plant, consequently in advance than blending it into your blend, soak cocopeat overnight in a bathtub or bucket with a hollow in it to wash away salts.

Perlite/vermiculite

Perlite is the decision of a truely happening mineral. A form of volcanic glass, created at the equal time as the volcanic obsidian glass gets saturated with water over a long time. Natural perlite is darkish black or grey colored amorphous glass. For use in gardening, the hard mineral glass desires to be processed right into a moderate, white color, that resembles styrofoam, To whole the transformation, overwhelmed perlite desires to be heated quick to 900 tiers Celsius.

The manner effects within the enlargement of the beaten portions of the mineral amongst 7 and 16 instances their specific period and amount, creating that light-weight tiny popcorn.

Advantages of using perlite for your potting combination:

1.Perlite maintains its shape even if pressed into the soil.

2.It has a independent pH level

3.It is non-poisonous and natural

four.It is high-quality porous and includes pockets of area inner for air

5.It can keep some amount of water promoting moisture inside the combination and while permitting the rest to drain away

Similar to perlite, Vermiculite is a hydrous phyllosilicate mineral that undergoes great growth at the equal time as heated.

Differences among perlite and vermiculite:

1.Both are obviously occurring mineral components which have among the identical tendencies.

2.Both enhance soil fine, particularly helping with aeration.

three.Perlite is known to be greater powerful at aerating soil.

4.Vermiculite is higher at keeping moisture.

5.Perlite is specifically much less high priced than vermiculite.

Neem cake powder

Neem cake is herbal manure is a bio product obtained within the system of cold pressing of neem tree fruit and kernels, and the solvent extraction technique for neem oil cake.

The AZADIRACHTIN content cloth fabric in neem cake herbal manure protects plant roots from severa fungi & pests and additionally works as a soil conditioner. Due to this Neem cakes were suggested to were (a) antifeedant (b) attractant (c) repellent (d) insecticide (e) nematicide (f) growth disruptor and (g) antimicrobial.

Neem cake manure performs the twin characteristic of every fertilizer and insecticides.

Nutrition Composition of Neem Cake Powder

Nutrient Element

%

Nitrogen

2.Zero% to 5.0%

Phosphorus

zero.50-1.Zero%

Potassium

1.25-2.Zero%

Calcium

0.5-3.Zero%

Magnesium

0.Three% to one.Zero%

Sulphur

0.2-three.Zero%

Zinc

15 ppm to 60 ppm

Copper

4 ppm to 20 ppm

Iron

500 ppm to 1200 ppm

Manganese

20 ppm to 60 ppm

Advantages of using Neem Cake Powder for your potting mixture:

1.Protects the basis from various fungal infections.

2.Neem cake improves the natural take into account content fabric of the soil, helping enhance soil texture, water retention functionality, and soil aeration for higher root improvement.

three.Neem cake manure can also reduce alkalinity in the soil with the resource of producing herbal acids even as mixed with the soil as a result ensure the fertility of the soil. It's a tremendous acidity enhancer for plants attempting acidic soils.

Bone meal powder

Bone meal is a combination of finely and coarsely floor animal bones and slaughterhouse waste products. Bone meal is typically used as a supply of phosphorus, calcium, and protein.

As a fertilizer, the N-P-K (Nitrogen-Phosphorus-Potassium) ratio of bone meal can variety notably, relying on the supply. From a low of 3-15-0 to as excessive as 2-22-zero. , even though some steamed bone food have N-P-Ks of one-thirteen-0.

IMPORTANT: According to recent Colorado State University studies, vegetation can quality get phosphorus from bone meal if the soil pH is underneath 7.Zero (acidic soil). That is due to the fact in alkaline soil excessive calcium content fabric cloth will bind to phosphorus and create calcium-phosphate this is unavailable to the plant. Therefore adjusting bone meal amount for potting mixture unique to plants is suitable.

Another factor to word is that while shopping for bone meal as powder, make sure that it is sterilized/steamed. This is crucial for proper storage and to avoid the improvement of fungus on the powder itself.

Mustard cake powder

Like Neem Cake, the mustard cake is also a byproduct received from mustard seeds after extracting the oil from the seeds. When mixed as an additive to potting combination, it provides to vital each macro-vitamins and micro-nutrients.

Nutrition Composition of Mustard Cake Powder

Nutrient Element

%

Nitrogen

2.Zero% to five.Zero%

Phosphorus

1.4%

Calcium

zero.6%

Magnesium

zero.6%

Manganese

zero.05%

Raw Protein

forty %

My potting combination recipe

Here's my tried and examined potting combination recipe.

30% Vermicompost

30% Cow Manure

20% Cocopeat (Washed)

10-15% Perlite/Vermiculite

10% Additive (Mix of neem cake powder + bone-meal powder + mustard cake powder)

Mix all of the materials and if possible positioned them in a sealed large barrel or bag and go away them to brew for approximately every week. This is optionally available but I truely have professional that it offers well effects.

IMPORTANT: While this recipe works for a massive kind of plant life I can't say that it's miles a normal mix. For example, some vegetation need greater moisture, at the same time as some don't. Adjustment in cocopeat and/or perlite is beneficial for unique plants.

My potting mix's texture

Soilless potting blend

How to degree proportions

Proportions are as clean as they sound but they may regularly confuse. Here are smooth strategies to get right accurate proportions:

1.By Weight: If you have were given a weighing tool, following a recipe turns into a exceptional deal easy for example 300g vermicompost, 300g cow manure, hundred-gram cocopeat, 100gram perlite, 30g+30g+30g (neem cake + bonemeal +

mustard cake powder). This may want to make a 1 kg potting mixture.

2.By length: Take a small container or mug now degree as 3 packing containers of vermicompost, three packing containers of cow manure, and so on. Now upload a handful of every additive neem cake powder and so forth in a unmarried area. You get the concept. Right? A bit of approximation proper right here however don't worry it'll paintings certainly notable.

to function bone meal or now not

When I first made this potting combo I added bone meal powder as aforementioned however as quickly as I had found out that bone-meal is excellent most effective when the potting blend is acidic, I stopped adding it to the number one mixture, as a substitute, as quickly as my mix have become organized and if I am developing any plant that loves acidic mix I upload bone meal to that.

A be conscious about dust/garden soil

"The soil is the notable connector of our lives, the deliver, and tour spot of all... Without right cope with it, we're capable of don't have any life"

—Wendell Berry creator Farmer

I am in no way in opposition to the usage of normal lawn soil for Terrace gardening however, besides the advantages of the soil-much less potting blend, I sense it is my sacred obligation to maintain soil. These are my motives for not using soil

1.We have abused soil with chemical fertilizers, industrial waste, and insecticides plenty that it's now not possible to discover natural natural soil.

2.It takes approximately 2 hundred years for 1 square centimeter of soil to shape so it's a nonrenewable beneficial useful resource.

three.I select out to go away the soil for the flora at the ground, they want it greater than we do. Billions of microorganisms assisting our environment relies upon on soil

By all manner, make a combination of your very own with the aid of manner of the use of lawn soil, you can lessen cocopeat, add extra sand or perlite and keep one of a type herbal depend like vermicompost & cow manure in same proportions as soil.

I request you to avoid the use of any chemical compounds- fertilizers, and insecticides in garden soil or in any other case. Your future generations will thanks for your noble deed.

Chapter 6: Secrets Of Feeding Your Plants

"Earth is aware about no desolation. She smells regeneration in the moist breath of decay."

Ask yourself one simple query. For how many days you may wait until you eat over again? We people consume multiples times a day for our our bodies and brains to art work maximum correctly and except there's a famine you can get your subsequent meal interior a few hours.

Lucky for us flora do now not want to be fed like this. Even luckier that plant meals want now not have cuisines as we do. It is constantly the equal vitamins for the plant however of route, the sources can be special.

One of the most famous questions I have become requested on social media after people see my flora is "What do you feed your plant?". It's moreover the maximum ignored difficulty of terrace gardening.

I am a sturdy practitioner of natural gardening and therefore this bankruptcy will talk about herbal feeding on your flowers.

Plant vitamins era

Let's take a small crash direction approximately the nutrients plant life want to live to inform the story, expand, bloom, produce end result, and so on.

The well-known n.P.Ok

Also referred to as primary macronutrients:

(N)Nitrogen: All plant life want it, without it plant life can't make plant protein that allows them in constantly constructing new tissues (read leaves, branches, and so on.)

(P)Phosphorous: Helps inside the manufacturing of flora and stop stop end result but most importantly a strong root gadget.

(K)Potassium: The energy powerhouse of the plant a.K.A carbohydrates which plants can only make with the help of potassium

Role of NPK

The now not so well-known

Also known as secondary macronutrients:

Calcium: Plants do no longer want heaps calcium however they really need sufficient to enhance the soil & bind the soil collectively.

Magnesium: Plants can't method daylight hours in the absence of magnesium. In brief, there may be no photosynthesis (that's how plant make their meals) if there may be no magnesium.

Sulfur: Like we people, plant life need proteins to construct & restore themselves and they produce plant protein excellent with the help of sulfur.

The thriller infantrymen

Also referred to as micronutrients:

1.Copper: vital for reproduction

2.Manganese: Works with plant enzymes to help wreck down carbohydrates and metabolize nitrogen

3.Iron: Essential for the formation of chlorophyll

four.Zinc: Regulates consumption of sugar inside the plant

5.Boron: Aids production of sugar and carbohydrates, chlorophyll synthesis

6.Molybdenum: Helps the plant seize nitrogen

7.Chlorine: Required for photosynthesis

In the previous bankruptcy, the natural elements of the potting aggregate cope with these vitamins and this may be maintained with normal feeding.

Feeding your plants

Plants are infants and now not like us people, they're babies for all time in terms of their food because of the truth they may be one hundred% depending on you to feed them and like toddlers, they are capable of only come up with signs. I can't don't forget if like infants flowers start crying or throwing tantrums. It might be a nightmare.

I researched, determined, and experimented with feeding my plants after making lots of mistakes. I'll confess as soon as I began Terrace gardening I hardly ever fed any of my plant life. Over time I figured that it isn't tough whilst you realise precisely what to do.

Secret 1: Regular feeding is the crucial component

The maximum important trouble to bear in mind is that your box has a limited surroundings and potential to hold nutrients for long. As the plant grows it consumes the necessary nutrients from potting blend and maintains on depleting it. Rains and water drainage also plays position in leeching the potting blend of nutrients. That's why plants need everyday feeding.

Implant this idea on your thoughts and you'll haven't any troubles feeding your flora.

Secret 2: Highest awesome potting blend

Recall from Chapter five flora don't care about their field till they begin to conflict to increase. The same is going for the potting combo. The potting mixture inside the box is the main supply of nutrients for the plant. Therefore a exquisite potting combination makes a top notch difference inside the manner your flora are fed.

A potting combination Is form of a powerhouse of herbal depend and slow-release fertilizers.

A properly potting combination considering that the start of a plant's journey guarantees that the plant keeps getting extremely good nutrients for max boom.

Secret three: foliar feeding

A nutritious potting combo looks after a majority of your plant's nutrition however you may be amazed to realise that flora may additionally soak up 8-20 times extra nutrients via their foliage than through the roots.

Feeding through foliar sprays is like people taking fitness dietary dietary supplements on top of normal meals to ensure maximum great fitness degrees.

Regular feed via foliar sprays gives your plant a health enhance and allows them hold its increase.

While natural fertilizers are slow-launch and consequently sluggish to absorb thru way of flora. Organic foliar fertilizers are fast-acting.

You can purchase formulated herbal foliar spray fertilizers from stores but in case you need to apprehend my desired ones, they are:

1.Liquid seaweed extract

2.Neem oil

There are 100s of natural spray fertilizers which can be available and loads of you may make your self but I propose you start with those   because of the fact they will be effective, monetary, outcomes to be had, and problem-unfastened.

7 commandments of foliar feed

1.Typically a foliar spray fertilizer is performed on the same time as the plant is transplanted, starts offevolved to bloom and primary give up end result begin showing. However, for a Terrace garden, I endorse using foliar feeding in uniform everyday durations (See Chapter 9. Secret To Fantastic Care for ordinary).

2.Spray your plants early morning whilst the climate is cool or middle of the night. Spray flora till you observe them dripping from the leaves (that is referred to as drenching). I opt to spray in the evenings round 6 pm so I don't need to worry approximately any form of leaf burn from sunlight hours.

three.If there can be a forecast for rain, then remove foliar spray to keep away from washing it away to waste.

four.Don't spray on the identical time because the winds are robust. The spray obtained't live on the leaves and couldn't be as effective. Also, due to the alternative wind course, you may get yourself in touch with the sprays, particularly to your face & eyes. It is higher to be steady than sorry.

five.Get a remarkable-notable spray pump. For some vegetation, a 1-liter hand spray will do. Personally use a 5-liter stress spray pump because of the reality I actually have quite some vegetation and it saves some of electricity and stops me from sore thumb and arms.

6.Make sure your flora are well watered the day you examine a foliar feed and the day after (especially on warmer days). This is critical for the nutrient uptake of the plant. Just keep away from overwatering.

7.The key to getting right consequences is a habitual for foliage spray.

Secret 4: rejuvenating potting blend

A potting combo may be depleted of vitamins through the years, that is a truth that maximum gardeners don't apprehend till their vegetation prevent developing or display bad symptoms of boom.

But why does potting aggregate get depleted of vitamins over the years?

1.Leeching of vitamins via drainage and non-stop rains.

2.Slow launch of natural depend thru drainage.

3.Plant's growth, age, and root gadget.

four.Heavy feeder flora

To recognize how vegetation burn up the potting combination, permit's apprehend that plants are:

1.Light feeders

2.Medium feeders

3.Heavy feeders

I would really like to apply the term patron in area of feeder for easy of records that flowers which may be are heavy clients use up the potting blend of nutrients faster than slight clients.

Just to recognize, below[6] are a few vegetable plants beneath top notch feeding education:

There are three procedures you may keep your potting blend fertile.

Top up approach

Just like a pay as you go cell recharge to hold the smartphone offerings walking, the top-up might be one of the terrific strategies to maintain your potting blend full of nutrients and fertile. How to try this:

1.Prepare a uniform aggregate of the subsequent devices inside the given proportions:

1.500 grams vermicompost

2.500 grams cow manure powder

3.a hundred grams neem cake powder

four.a hundred grams mustard cake powder

5.one hundred grams bone meal powder (simplest for acidic potting blend)

6.OPTIONAL: a hundred grams of timber ash

7.OPTIONAL: Handful of perlite, it acquired't harm to function a bit of drainage and aeration but it is not essential.

8.OPTIONAL: Handful of cocopeat in case your plant wants to keep extra water.

2.When you're plant is due for watering, spread a 1-2 inch layer of this aggregate on the floor of potting blend for your area.

three.Water slowly first only so that the top blend is moist. After you end watering all boxes with this addition. Then water once more from the primary container but this time do deep watering till the water begins offevolved offevolved draining out. This is not important to do however in my revel in, I actually have located that the first round of water settles the combination nicely on the floor certainly so even as you water once more, it receives disbursed uniformly.

4.This method works first-rate for perennials and fruit bushes.

five.It takes everywhere amongst a fortnight or month for natural fertilizers to break down and start providing nutrients to plant therefore that is a fantastic gradual-release feeding technique.

6.This approach also enables in mimicking the natural atmosphere in forests in which layers of herbal rely upon the floor are normal over years, proving home to beneficial fungi, bacteria, and food for the plant.

In the subsequent financial disaster, I will proportion my habitual approximately the frequency of this approach.

Rejuvenate on the equal time as repotting

When your plant outgrows your container or if the sector is broken or at the same time as your plant is suffering for space to increase then it's time to repot your plant.

Repotting is a high-quality opportunity to make up for the loss of nutrients inside the soil.

There are methods to re-pot:

1.Repotting with the whole pot ball of the plant: In this method, you extract the plant with the whole combo that has everyday the form of the box over the years. This is typically finished even as you probably did no longer anticipate the plant to outgrow the pot so fast.

1.Make superb the plant is not watered for a day or two.

2.Extract the plant lightly by way of tapping the edges and loosening the combination.

3.Fill the latest container with the smooth potting combo, press it to clean air wallet

four.Put the whole ball in order that the floor of the plant fits the floor of the latest box.

5.Fill potting combo throughout all sides urgent firmly.

6.Water thoroughly.

7.Keep in partial color for an afternoon or .

2.Repotting with root pruning and trimming: In this technique as the call indicates, you prune the extra roots (typically the skinny fiber-like from the sides) from the plant.

1.The steps are similar as aforementioned however as quickly due to the fact the plant is extracted, lightly dirt off the potting aggregate from all aspects to show skinny fibrous roots.

2.Then the usage of a pointy sterilized scissor trim the roots. To reduce the scale of the ball. Don't do it excessively except you advocate to make a bonsai out of the winning plant.

3.If roots are outgrowing the vintage subject from the bottom, you may chop them off too.

four.I opt for this approach once I purchase plant life from the nursery (without root pruning). My purpose is to put off the clay soil as loads as possible.

5.Now prepare a liquid combination of 5ml liquid Humic acid + 1 ml of liquid seaweed extract in 1 to one.Five liter of water and drench the plant after you end such as the potting aggregate. This will assist the plant get better the surprise, promote quicker root boom, and moreover promotes microbial activity inside the combination.

Rejuvenate antique potting combination

Potting blend is a precious commodity for your Terrace garden. When you have an entire lot of flowers this may in reality upload to your prices besides you are making your compost, getting cow manure with out price. But whether or not you have got had been given a zero-fee variety lawn, it's far a outstanding concept to reuse vintage potting mixture after rejuvenating it. But allow can help you know that that is a tedious procedure in spite of the fact that the cease end result is profitable.

Before we communicate approximately rejuvenating potting combination, allow's recognize a way to pick out a potting combination for reuse:

1.If your plant suffered via bacterial or fungi infection or modified into stricken by insects in some way, then I may no longer recommend it for reuse. However you still need to apply it, you may must use it thru bug composting[7] according to analyze in Cornell University (NY USA)

2.I might no longer recommend reusing potting combo if you have grown heavy feeders very last year. Such potting mixture has a better diploma of vitamins depletion.

3.Check for Ph tiers, See Chapter 8. Basics of Potting Mix Acidity (PH)

four.This works awesome for veggies and kitchen gardens.

How to rejuvenate vintage potting combination: Here is the grade by grade way to rejuvenate your vintage potting mixture

1.You might need a large plastic sheet, a massive field with drainage holes, a trowel, and a rake if feasible. Large plastic luggage.

2.Spread your blend on the plastic sheet, a rake is available in available right right here. Remove portions of roots any unwanted stones and so on. Loosen any clumps that would have shaped.

3.Wash your blend: Take a huge barrel with drainage holes, blanketed with a plastic net. Fill it with soil and saturate it with water.

4.Once the water stops draining, pour the potting mixture returned to a plastic sheet, unfold it, and leave it within the solar till certainly dry and fluffy

five.50/50: Once the combination is dried, take an equal quantity of the clean potting mixture and mix them each.

6.OPTIONAL: To adjust Ph upload Neem cake or bone meal powder or lime SeeChapter 8. Basics of Potting Mix Acidity (PH) for added information.

7.Put this combination in baggage and leave them in dark & dry storage for about 2 weeks to remedy and brew.

8.That's it your rejuvenated potting mixture is prepared.

9.When you use it to plant, don't forget about approximately to feature the Humic + Seaweed mixture said while repotting.

You can also try this technique 2 instances in your antique mixture and in some times likely 3. After that, your mixture loses its capability to recharge itself.

Make your herbal fertilizers

There are 100s of natural fertilizers in the form of foliar sprays, potting mix amendments, and masses of others. Are to be had these days commercially that makes it very smooth to feed your plant life in a whole lot of techniques. So why could you need to make your natural fertilizers:

1.As your garden grows so does the want for fertilizers expand and that could get costly through the years or require a devoted finances (Honestly you'll however need a charge range)

2.Purity and best of fertilizers: Like a few different commodity cheap fertilizers also can be adulterated with dangerous chemical substances.

three.It is usually pinnacle to have alternatives. During the Covid-19 lockdown, I had no preserve-sold herbal fertilizers and that's even as my home made fertilizers got here to rescue

So except home composting, there are various sorts of fertilizers that you could make your self however the following are a number of the simple, tried, and tested organic fertilizers recipes:

Moringa Foliar Spray

Moringa (Sehjan in Hindi) has currently been accumulating so much hobby because of its extremely good fitness benefits for humans. Flowers, leaves, and culmination all are suitable for ingesting and feature terrific dietary values as a end result making it a superfood for human beings.

You can discover Moringa wood almost anywhere in India at homes, societies, roadside, highways, or even in forests and that makes the supply of moringa leaves incredible easy.

We should make an terrific foliar spray fertilizer (a plant boom promoter) for our flora the usage of Moringa leaves. This is almost a unfastened fertilizer for your Terrace lawn.

A examine posted in 2000[8] suggests that Juice from easy moringa leaves can be used to deliver an powerful plant increase hormone, growing yields with the beneficial aid of 25-30% for nearly any crop. One of the energetic materials is Zeatin: a plant hormone from the Cytokinines business enterprise.

Here is the particular recipe from the observe on the way to make the spray:

1.Make an extract by using the use of way of grinding younger moringa shoots (leaves) (not extra than 40 days antique) together with a bit of water (about one liter in step with 10 kg sparkling fabric).

2.Filter the sturdy out of the solution. This may be finished through setting the answer in a cloth and wringing out the liquid. The robust bear in mind, with a purpose to contain 12-14% protein, can be used as cattle feed or can be put in compost making.

three.Dilute the extract with water at a 1:32 ratio and spray right now onto flora (if the extract isn't going for use indoors 5 hours, it's far excellent saved in a freezer until wished). Apply approximately 25 ml in line with plant.

4.The foliar spray have to be implemented 10 days after the number one shoots emerge from the soil, yet again approximately 30 days in advance than plants begin to flower, over again whilst seed seems, and in the end another time inside the path of the maturation segment.

For box-grown plants, you could spray this on weekly foundation for extremely good increase.

Another opportunity approach is to boil the Moringa leaves in water and stress the leaves and spray this liquid for your vegetation on weekly basis. Take approximately five to ten kg of leaves and boil them in water. The shelf life of this technique is longer as you can keep the liquid for few weeks. I recommend you are making it on monthly basis.

Cow dung and cow urine foliar spray

While searching out a low-price do-it-your self herbal fertilizer made the use of cow dung. This is a totally smooth recipe but it outcomes in a fantastic natural fertilizer. Here is the recipe.

1.Things you could want: An earthen pot (matka), A 20-liter paint bucket, a piece of cotton string about a feet prolonged, 10 kg smooth cow dung, 10 kg cow urine, about 1kg of neem leaves

2.Make a small hollow in the earthen pot, about the thickness of a pencil, allow the string skip via it, tie a knot at the give up simply so it continues placing through the hollow.

3.Mix cow urine and dung well. Make neem leaves paste thru grinding the leaves and mix them.

four.Place the earthen pot above the bucket and transfer the aggregate to the pot. Cover the pot starting with some thing.

5.Over a while there acquired't be lots liquid left in the earthen pot, that is in which you switch the collected liquid in a few air-tight plastic bottle and you could save it from 6 to 3 hundred and sixty five days. Also, you may add greater cow urine to the combination and allow the manner begin all all yet again.

6.For spray dilute this aggregate in the ratio of 1:10 elements of water and spray weekly.

Other herbal fertilizers which you DIY

There are dozens of herbal fertilizers (foliar spray, increase promotors, microbial interest enhancer, and soil enhancer) that natural gardening specialists have advanced and they will be pretty famous. You can also additionally search for them at the net. Here are a few:

1.Jeevaamrut

2.Panchagavya

three.Amudham

4.and masses of extra.

There is an outstanding useful resource about making the aforementioned fertilizers and loads of others. It's the Soil Recipes (Hindi & English) ebook by the use of Isa Alvares[9] (The Organic Farming Association of India). You can down load this e book without a doubt unfastened from their internet web website. At the time of writing the ebook, the unfastened link given inside the footnote changed into live and strolling

Waste decomposer – your first-class pal

National Center of Organic Farming (NCOF) in Ghaziabad has made a product named Waste Decomposer that can be used for quick composting from herbal waste, soil fitness development & as a plant protection agent. It is a consortium of few useful microorganisms this is remoted with the aid of Dr. Krishan Chandra from Desi Cow Dung and took 11 years to standardize the mass multiplication method at the farm degree. Waste decomposer works as Biofertiliser, Biocontrol, and similarly to Soil Health Reviver.

In quick due to the fact the call advocate waste decomposer is a way of lifestyles, it's miles a consortium of microorganism extracted from desi cow dung. It may be used for quick composting from herbal waste like leaves, stays of roots and stalk after harvesting, culmination, flowers, fruit &

vegetable peel, cow dung, agriculture waste like chaff, and so on.

Although this product is made for farmers, you don't hesitate to use it in your Terrace lawn because it has were given exquisite blessings and it can artwork wonders in your plants.

Here's the NCOF recipe to make a waste decomposer solution, this solution is the concept of all feasible usages of the product.

1.Take 2 kg jaggery and combined it in a plastic drum containing hundred liters of water.

2.Now take 1 bottle of waste decomposer and pour all its contents in a plastic drum containing jaggery answer. Avoid direct touch of contents with fingers.

three.Mix it properly with a timber stick for uniform distribution of waste decomposer in a drum.

four.Cover the drum with a chunk of paper or cardboard and stir it every day a few instances for three-4 days.

5.After 5-7 days the solution of the drum turns creamy/dwindled yellow.

Waste decomposer answer on my terrace

Once the above mixture is ready you may by no means have to buy the waste decomposer bottle yet again as you could put together the waste decomposer solution over and over from this answer. For this, 20 liters of waste decomposer solution is brought to a drum with 2 kg of jaggery, and 20-liter water is added.

If you preserve on saving 20 liters of this answer, then you can have a waste decomposer answer on your lifetime.

Using waste decomposer on your terrace lawn

I want waste decomposer modified into there a decade within the past. As a terrace gardener, this could make your lifestyles so smooth, and with this product, all your belief structures approximately saving the environment, reuse, and recycle will come into motion. So applicable and easy.

There are many ways you may use waste decomposers on your terrace garden however due to the fact this financial ruin is all approximately feeding your plants, the following are the do-it-yourself fertilizers that you can make at domestic using waste-decomposer:

Waste decomposer answer recipe for terrace garden: Take a 20-liter plastic bucket and positioned 10-liter water, 100-gram jaggery, 1-2 teaspoonful of waste decomposer, cover it with a cotton fabric, and go away it in color for 7 days however stir the combination instances every day with a wooden stick.

After 7 days your waste decomposer solution is ready to apply as follows:

Micronutrients foliar spray

Ingredients:

1.Moong dal 100gm high-quality powder

2.Tuar dal 100gm fantastic powder

3.Chana dal 100gm terrific powder

4.Udad dal 100gm satisfactory powder

5.Mustard seeds ground or mustard cake powder 100gm first rate powder

6.White or black til (sesame) 100gm quality powder

7.Sunflower seeds or any oilseed 100gm exquisite powder

8.four-5 iron nail

nine.A small copper glass or utensil or copper wire 100g

10. Geru Mitti 50 g (OPTIONAL in case you don't have)

Method:

In a plastic bucket take the waste decomposer answer and upload all of the components above. Stir to mix properly the use of a timber stick and go away it for 10-15 days protective with a cotton cloth. The cease result may be a thick solution that you could keep in an airtight field. This solution can last up to 3 hundred and sixty 5 days.

Application:

Mix a hundred ml of this answer in 1-liter water and do a foliar spray. You may additionally spray this on a weekly or biweekly foundation in a few unspecified time inside the destiny of the developing season, specifically in advance than flowering and fruiting.

Liquid manure using kitchen waste

Ingredients:

1.Wet kitchen waste like vegetable and fruit peels, vegetation, and so on.

2.About 1 detail jaggery in comparison with the amount of above waste.

three.Waste decomposer answer

Method:

Take a plastic bucket and fill all your kitchen waste and jaggery proportionately. Then upload a waste decomposer answer so that all the kitchen waste is soaked. Cover it with a cotton material. Now preserve inclusive of kitchen waste each day and waste decomposer solution if required to soak similarly. Once the bucket is complete, depart the combination for 10 days and restart the machine in every one of a kind bucket. After 10 days the use of a cloth gather the liquid.

Application:

Take 1 a part of this liquid with 10 additives of water or 100 ml liquid with 1 liter of water. Apply this solution straight away on your area's potting blend weekly or bi-weekly.

Moringa foliar spray

Ingredients:

1.Fresh inexperienced Moringa leaves approximately 5kg

2.Waste decomposer answer 20 liter

Method:

Grind moringa leaves to make a quality green paste, upload little water if desired for grinding. In a plastic difficulty with a cover or bucket, blend moringa green paste and waste decomposer solution using a wooden stick.

Cover the answer and go away it for 7-10 days. During this era stir the answer properly the use of a wood stick 2 instances a day.

After 10 days the usage of a fabric or sieve, accumulate the liquid and maintain it in an air-tight bottle.

The solid detail can be in addition located into the kitchen waste decomposition bucket or can be used as a soil change.

Application:

Take 1 a part of this liquid with 10 factors of water or one hundred ml liquid with 1 liter of water. Spray over flora weekly or bi-weekly.

Now you need to be thinking that I may additionally rather use the moringa foliar spray cited in advance on this financial disaster. You genuinely can however this one will provide more advantages due to the fact the microbial hobby in waste decomposer will permit quicker absorption of the vitamins from the moringa leaves.

If you desire to have a look at greater approximately making your herbal fertilizers & pesticides, I pretty suggest you to study Jaivik Kheti adequate Nuskhe - Venkatesh Narayan Singh.

Chapter 7: Organic Pest Control

"Most of the time pests and infection are truly nature's manner of telling the farmer he's performing some issue incorrect."

When I began gardening I didn't even apprehend about pests, as a minimum now not till they created havoc on my plants now not as soon as but typically. Over years I had found out that pest control is a everyday, no longer a scenario that may occur occasionally out of nowhere that you need to cope with.

Let me display you a number of the maximum realistic hints of pest manipulate that I had observed and experimented with over years and the manner they helped my plant live healthful. I will simplest be sharing herbal pest control.

But in advance than I try this surely take a 2d to anticipate why some of us are wholesome at the same time as others have loads of fitness troubles. Because the healthful group is aware of SELF-CARE and it is a ritual, a routine, and an act of kindness to yourself.

You want to expose the equal deal with your flora.

Daily ritual

You have a few plant life, they're all wholesome and blooming or flowering but one top notch day, you observed that a number of your flora have started out to appearance unhappy, some has curled leaves or having cuts, and so forth at the same time as you had a near appearance you decided out that it is infested with a pest. There are in all likelihood subjects you may do to embody this case:

1.Rush to spraying chemical insecticides because saving your plant is your top precedence.

2.The infestation has grown beyond control and you haven't any opportunity however to amputate your plant's infected areas and want that it's going to get better.

If the number one choice works, probabilities are your plant is saved but your vegetation, its potting combination might also additionally have been contaminated with the aid of the chemical materials that could harm the overall surroundings of the plant, and worst if it's an stable to consume plant, all the chemical compounds will go inside your body.

"An apple a day may additionally have stored the medical physician away in advance than the industrialization of food growing and schooling. But, in accordance to analyze compiled with the aid of using manner of the United States Drug Administration (USDA) in recent times's apple includes residue of 11 remarkable neurotoxins—azinphos, methyl hloripyrifos, diazinon, dimethoate, ethion, omthoate, parathion, parathion methyl, phosalone, and phosmet — and the USDA became trying out for high-quality one class of chemical materials called organophosphate insecticides. That doesn't sound too appetizing, does it? The average apple is sprayed with insecticides seventeen instances in advance than it's far harvested."— Michelle Schoffro Cook, The Brain Wash: A Powerful,

All-Naural Program to Protect Your Brain Against Alzheimer's, Depression, Parkinson's and Other Brain Diseases

You apprehend why we ought to do herbal terrace gardening. A clean but very effective every day ritual is of essential importance for pest manage, relying on the size of your terrace garden, this ritual may additionally take from 15 mins to at the least one hour or greater for a large terrace garden:

Inspect leaves/give up end result/plant life

Things to test for your plant's leaves:

1.Color: Check if the shade of the leaf is unusual or if there are any uncommon spots. This is often a signal of nutrition deficiency but also can be a pest hassle.

2.Texture: if leaves have a few form of powdery substance on pinnacle probabilities are it's a powdery mould

3.Leaf curling: Pests like aphids, thrips, mites, and whiteflies purpose leaf curl on pepper flowers.

4.New shoot and leaves: Most pests prey on new plant growth

five.The underside of the leaves: This may be very important as maximum pests conceal under

6.Check the buds for any discoloration: Usually a nutrients problem but also can be a pest hassle

You might be capable of manipulate a majority of pests sincerely with the useful resource of the usage of reading your plant leaves each day. I keep a small bucket of water blended with neem oil or definitely cleaning cleaning soap-like liquid handwash and handpick any recognized pests drop them into this bucket. Trust me, this may prevent from pretty some hassle in the future.

CAUTION: Don't touch wholesome plants or healthful parts of vegetation with the same hand you used to select insects as it is able to encompass eggs and may infect specific flowers.

Some pests are too remarkable at hiding and nature has given them such colours that they may combination themselves with stems and trunk of the plant to make a definitely perfect hideout. For instance mealy insects, a noticeably healthful-looking stem also can have a whole colony of mealy insects beneath.

Inspect dropping of healthful leaves

If a fantastically wholesome-looking plant begins dropping leaves with none reason like a natural fall, it is able to be a fungal contamination within the root. This is probably a extreme trouble if left unattended. Check:

1.If the potting is mixture soggy, generally due to horrible drainage.

2.If the drainage is blocked

three.Green algae formation at the ground is an indication of water taking too prolonged to empty from the world

Note: Good drainage in a pot and a nicely-tired potting mixture can lessen the above issues to a brilliant quantity.

Try the following to improve the situation:

1.Stop watering the plant for an afternoon or so and phrase if the plant recovers.

2.Take a handful of neem cake powder and unfold it on the sector floor near the trunk, water lightly.

three.If you have large drainage holes as I sincerely have described in Chapter four. Secrets To Choosing The Right Container, tilt your area, take an vintage toothbrush and scrub the plastic net to clean the blockage of holes.

four.If now not one of the above works, then repotting is your only desire till you decide to apply a few chemical fungicide which I could not advocate.

Weekly ritual

I'm luckily dealing with my terrace lawn which has ornamental, flowering, greens, and culmination plant with this weekly ritual. It is extraordinary smooth and powerful but situation is the crucial component to its success.

Every weekend on Saturday at five:30 pm, I put on my lawn apron, fill my 5-liter sprayer and drench all my flowers with this miracle herbal pesticide plus fertilizer. Yup, your wager is proper, it's the Neem Oil. If you notice the tables below, you may recognize why Neem oil.

From the table above it is clean that neem oil makes your plant proof toward pests in lots of approaches, Neem oil show off the same homes as neem cake viz. (a) antifeedant (b) attractant (c) repellent (d) insecticide (e) nematicide (f) growth disruptor and (g) antimicrobial.

What all do you need to prepare the Neem Oil Pesticide Spray?

1.Pure neem oil: Pure neem oil solidifies at a less warm temperature like coconut oil. With the boom in cognizance about neem oil, business enterprise makers are promoting this at a much better charge. Try to supply it locally, government. Natural departments promoting neem oil and distinct organic upload ons and so on.

2.Agriculture grade top notch Neem oil: If pure neem oil isn't always to be had. This is usually to be had thru the call of Neem oil or Neem Oil Concentrate or AZADIRACHTIN. Besides domestically sourced natural Neem oill, I have used the following and they all have proven the identical effects:

OrganicDew Neem Oil Concentrate

AZA Power+ from IFFCO: This one is emulsified.

three.Water

four.Some natural handwash liquid simplest in case you are the use of herbal neem oil

five.A spray pump: hand or pressure pump

Neem Oil Pesticide Recipe:

1.Take 5ml of Neem oil

2.OPTIONAL: Take 1 ml of liquid soap ONLY if natural neem oil is used.

three.Mix 1 neem oil (cleaning cleansing soap liquid if required) with 1 liter of water. You can increase the amount proportionately for instance 25 ml neem oil for 5 liters of water.

4.Mix thru shaking the pump field carefully.

5.Your spray is prepared

Neem oil Pesticide Application:

1.Spray sooner or later of midnight hours, I spray amongst five-6 pm, usually after the sunsets.

2.Sometimes mainly all through summer time, the times are heat round 5-6 pm, if feasible spray later. Hot climate can cause leaf burn.

3.While spraying, drench the complete plant, stem leaves, and beneath the leaves.

4.If you feel rain then delay the spray to day after today because of the reality spraying at the same time as the rain is close to will wipe all of the liquid from the flowers.

5.Similarly, keep away from spraying on a windy day.

This weekly dependancy will defend your lawn from a majority of pests. The secret is subject.

What if I pass over the weekly ritual

It is ok, every now and then you're in reality not to be had or has some different priorities that you could not discover time to carry out this ritual. So what are you capable of do? It's smooth, do it the following day or the following after that. But make sure that you do it and don't pass it altogether for the week.

Damage control

We all make errors and perhaps you disregarded seeing a likely sign of infestation. What to do?

1.Start the spray as fast as you test the number one symptoms and signs of infestation. If the infestation unfold it can take longer to treatment. You can also additionally need to reduce off inflamed factors or worst use chemical fertilizers, which I should not suggest.

2.Repeat the remedy after 3-four days until you note development and infestation is lengthy past.

Beneficial insects

If you need to have a few guards doing pest manage for you on the equal time as you are a protracted manner from your flora, that's in which the ones beneficial insects come. They are usually ladybugs, lacewings, predatory stinkbugs, soldier beetles, tachinid flies, parasitic wasps, and plenty of extra.

These groups of insects are called predators and parasitoids. They can devour heaps of pests every day.

This is an extended-term task to attract those useful insects for your lawn this is beyond the scope of this book, consequently I encourage you to look at Attracting Beneficial Bugs to Your Garden By Jessica Walliser. It is a exquisite ebook to recognize and enforce natural pest manipulate to your lawn.

Making natural insecticides at domestic

I keep in mind neem oil to be the high-quality natural pest manipulate in your terrace lawn if finished proper however there are a few distinctive natural do-it-your self insecticides that I determined to be pretty useful as properly.

Neem & waste decomposer pesticide

A pretty smooth however effective recipe the usage of neem leaves and waste decomposer answer that I actually have defined inside the preceding economic break

Ingredients

1.Neem leaves grind to make ½ kg paste

2.4.Five lit WD solution

three.5-liter water

four.A 10-liter plastic bucket.

Mix all of the above thoroughly the use of a timber stick and depart the mixture for three weeks. You can also stir the mixture instances each day however no longer essential.

Application: Mix 50 ml of the combination in 1-liter water and spray to drench the plant on weekly basis or at the same time as there may be any sign of infestation. For much less hard spray utility, sieve the solution earlier than mixing it in water.

The masala pesticide

I call this one a masala pesticide because of the truth this is be made using common kitchen vegetables commonly used to make sabzi masala.

Ingredients

1.Onions

2.Ginger Root

3.Garlic

4.Hot green chili

five.OPTIONAL: Neem leaves

6.Some herbal liquid cleaning cleaning soap or hand-wash liquid

Take all the above substances in equal quantities or crush them in mortar & pestle. Boil the components to make a decoction, sieve it, and shop it in a plastic bottle.

Note: Instead of making a decoction you could additionally make it the usage of waste decomposer solution like neem leaves.

Application: Mix one hundred ml of decoction, five-10 ml liquid cleaning soap in 1-liter water and spray to drench the plant on weekly foundation or whilst there may be any sign of infestation.

The sanitizer pesticide

Due to the Covid-19 pandemic rubbing alcohol, moreover referred to as alcohol sanitizer may be very effects available in recent times. Keep a small spray bottle usually accessible on your garden as this one allows in controlling one particular pest, the mealybugs, and the exceptional trouble is that it does no longer damage your plant. When you observe a bug, spray it near and soak the laptop virus until it will become slight brown, it's an instance that alcohol has penetrated their pores and pores and skin and they're nice to die.

Remember that could be a prevention approach, now not a way to control a complete-scale infestation.

Karanj oil

Similar to Neem oil, Karanja oil is a bio-controlling agent that acts as a repellent, preventive, and healing agent in opposition to pests and bugs. Prevents the development of insects the least bit tiers - person, larvae, pupae, and egg and it has a totally particular character of the ovicidal hobby.

Karanj oil is concept to be effective in competition to a huge massive type of insect pests. Karanj seed oil is concept to encompass lively metabolites consisting of karanjin, pongamol, glabrinand pinnatin, and so forth. Karanjin is powerful in competition to a large large shape of bugs. It additionally has better antifungal homes.

You might also additionally change your weekly neem spray with karanj spray for vast spectrum safety in the direction of pests, bugs, and fungal ailments.

Prevention is…

…better than treatment. If you take into account this suitable old announcing, you apprehend exactly what it way. With the development in farming and pest manage strategies, there are a few powerful methods to save you pests and illnesses to affect your garden at the deliver diploma.

These strategies are quite effective in that you may save you about 50%-60% of pests & ailments to ever touch your garden if used appropriately. I even have used such strategies and that they have tested to be very useful.

Fruit fly lure

Fruits flies are chargeable for infesting your garden with an entire lot of pests and ailments. For example, if you have ever visible a maggot inner a fruit, recognize that fruit flies are responsible for that.

A fruit fly entice is a wonderful and strong way to reduce the variety of fruit flies. This method makes use of a pheromone (a completely particular fragrance that lady flies produce and the male receives interested by it) to attract the male fruit flies proper right into a lure in which they're killed. Even even though great the male flies are killed, this makes it extra tough for the girls to find out a mate to make extra maggots.

They are a reasonably-priced, chemical-unfastened, and solid tool for pest manipulate specifically if you are growing veggies and fruits.

Sticky lure

Yellow coloration & blue color is natural attractive shades for small sap-feeding bugs like whiteflies, thrips, aphids, leafminers, jassids, fungus gnats Insects are attracted and attempt to take a seat down on the trap and due to the gum, they stuck and die in short time

These traps are generally non-poisonous, prolonged-lasting, whether or not or not-evidence, and inexperienced.

Customize your pest manage routine to fit you should you come across pest troubles an lousy lot a great deal less often, as an example, you could do neem oil spray as soon as in 15 days.

Some gardens are greater at risk of pests at the same time as others are much less. This is predicated upon on loads of factors however you may make sure of one problem that during which there is a plant possibilities are there may be pests sooner or later.

Chapter 8: Basics of Potting Mix Acidity (PH)

"Went searching out religion on the wooded region ground, and it confirmed up everywhere. In the solar, and the water, and the falling leaves, the falling leaves of time."

Usually, you do now not need to fear about the PH balance of your potting combo and that's a superb factor due to the reality unnecessarily messing up with the ph balance may furthermore have an impact on the overall fitness of your plant in a awful manner or it can even kill the plant.

What is ph?

In layman's phrases ph diploma determines the acidity of potting combination. The pH of potting combo determines the absorption of vitamins via the flora.

0 Most Acidic 7 Neutral 14 Most Basic

The plant prospers on the same time as pH is among 5.Zero and eight.5. If the ph is past this variety in both course, the problem occurs. When pH values are decrease than five.Zero, manganese & aluminum can emerge as poisonous to flowers. Plants get more of these liberated nutrients and cannot manner them, leading to plant loss of existence.

Signs of low ph

1.Yellow spots and results in browning and leaf demise.

2.Wilting leaves

3.Stunted growth

four.Blighted leaf suggestions

5.Yellowing of foliage

6.Other leaf discoloration

7.Poor stem improvement.

WARNING: Please recognize that the above symptoms can also be due to unique factors like dietary deficiency, lack of light, incorrect watering, and so on. So earlier than you begin strolling on figuring out and fixing the Ph, do check those elements.

So here's my top advice, don't worry approximately the PH balance of your potting blend besides the entirety else fails.

What motives low PH

Several factors drop the ph degree of your potting mixture and render it acidic.

1. Rainfall: Naturally acidic however it has a lot much less effect in town areas so consider yourself fortunate if you live in a metropolis. But this does not simply arise in a single moist season.

2. Chemical Fertilizers: Especially nitrogen fertilizers that include ammonia. The largest perpetrator.

3. Organic rely: Yes, that's right, decomposition of herbal rely makes the soil acidic. Always use nicely-decomposed manures.

four. Municipal Water: Treated water may increase the acidity of your potting mixture.

How to check PH diploma?

You can resultseasily take a look at your soil ph through using using an effects to be had tool referred to as a PH meter, it's now not expensive and might prevent from an entire lot of hassles. You can monthly check your potting combination's ph level if the aforementioned factors are examined thru you.

A ph meter

How to restore PH?

In my enjoy, a top notch potting mixture, properly timed addition of macro and micronutrients, proper watering, and slight can prevent from PH-related troubles. But if no matter doing everything right your plants aren't thriving, please do test your potting blend ph and connect it.

Method 1:

Always use well-decomposed cow manure, vermicompost, and compost. This resists changes to potting mix PH.

Method 2:

If you stay in regions wherein moist seasons are non-save you and longer, offer a obvious/translucent cover/color to your Terrace lawn in order that they get sufficient mild however are protected from extra rainwater.

Method 3:

Add Lime (Chuna in Hindi). If you stumble upon a low ph then including lime to your combination allows enhance the ph levels. Loosen the pinnacle mixture thru gently hoeing for approximately 4-6 inches and mix the lime powder. Start which includes in small portions first and retake the studying after 3-four days.

Method four:

When your soil ph stage is higher that means it's more alkaline. Sulfur permits in bringing down the pH degree. Add small quantities of sulfur considering the reality that a larger quantity also can kill the plant.

This have turn out to be a tiny crash direction about ph diploma of the potting combination. I must propose similarly exploration of the issue in case you enjoy it as a main trouble for your terrace garden.

Chapter 9: Secret To Fantastic Care

"I do no longer maintain my garden. I maintain my sanity with the aid of maintaining my garden."

When you adore and deal with your terrace garden, you connect a nicely-defined, systematic set of sports sports and a well-defined regular. In this financial catastrophe, I will percentage my regular for the fantastic care of your terrace lawn which took me years to determine out. A properly-defined recurring on your garden furthermore helps in green time management for precise crucial matters on your lifestyles.

Daily ordinary

Don't be overwhelmed through the sports activities that you have to do each day on your lawn as they're quite simple, clean to do, and feature many advantages in long term.

Daily 1: Inspection of plant

As I even have defined in depth approximately the every day inspection of your flowers, please speak with Chapter 7. Organic Pest Control

Daily 2: Removal of antique/yellow leaves

You won't find vintage leaves to your lawn each day if you hold getting rid of them. Although they will fall on their personal in a while so this hobby satisfactory has aesthetic charge to your lawn.

Daily three: Removal of fallen leaves

Plants mainly with smaller leaf sizes can also fall and get gathered at the field ground. While you might imagine that leaving fallen leaves to decompose also can move lower back treasured vitamins to the soil, presents habitat for loads of crucial and treasured insect species over winter, and acts as a natural mulch. Right? No

Why? Because rotting natural rely produces acid (Remember Chapter eight, what causes low ph), the more of which may additionally additionally damage the plant mainly to the sensitive one. Also, It Is able to become home to ants and different no longer-so-useful insects that could motive a nuisance.

Rotten leaves

So my belief might be to hold a separate pot/bag for collecting fallen leaves as they may be an super deliver of organic don't forget. You may moreover additionally located them in your house composter if any or use them with waste decomposer solution or you can furthermore use dry leaves as mulch.

Daily four: Conscious watering

The most apparent one but it is able to be advanced. Remember you don't need to water every plant every day till off direction you live in a hotter vicinity.

1.Know approximately your plant's precise water requirement.

2.Follow the finger rule to test in case your vegetation need watering. Press your finger within the subject combination for one inch and if it feels strive, your plant might also additionally moreover need watering. With time you can understand watering desires for precise plants truly by way of manner of searching at them.

3.Always try to water the usage of a watering can with a bath This permits saving water and additionally lets in in spreading it evenly. It furthermore maintains the top floor undisturbed.

four.Water slowly.

Effect of watering right away the usage of a pipe or a mug

Daily four: De-weeding

Remember your plant has a restricted surroundings and you don't need weeds to take over masses that your plant has to compete with them for vitamins. Therefore take away weeds when they seem first, especially in moist motives. Over a while this interest may not require every day efforts.

Weekly routine

I keep all my hard work-massive gardening responsibilities for weekends. You can pick out any day during the week. I select out weekends to take a break from my standard paintings. As loads as weekly sports activities are vital to your lawn, they're similarly useful for you as properly. Before I speak approximately garden-related sports activities, allow me show what blessings I get out of my weekly gardening sports:

1.It permits me unwind. In easy terms, it's amusing in its way.

2.It's a digital detox for me as I am some distance from my phone and pc a majority of the time once I'm strolling on my lawn.

three.It promotes bodily interest.

four.I get a experience of accomplishment as soon as my gardening obligations are executed.

five.I furthermore get a feel of fulfillment for my responsibilities in the path of my plant toddlers similar to I get for my youngsters.

6.Chilling out with friends or family in the night with beers and scrumptious food feels profitable after this form of day.

I think I should stop in advance than this can emerge as like a '10 detail to kill your pressure' shape of financial ruin ;-)

Weekly 1: Neem/Karanj Oil Spray

This one interest has saved me from pests growing havoc on my terrace garden. Behind my residence is an empty plot and it becomes a jungle in some unspecified time within the future of rains and remains like that until summer season. While it's miles a deal with to the eyes to awaken to such greenery however it moreover draws a whole lot of insects, pests, and whatnot.

Every Saturday amongst five:30-6:30 pm (relying on sundown) I spray neem oil on all of my vegetation. I actually have defined in more depth in Chapter 7. Organic Pest Control .

Do this hobby religiously and your garden can also have just a few pest issues.

Weekly 2: Labor-extensive duties on weekends

If you are already preserving your garden on each day basis there received't be many difficult paintings-incredible tasks for the weekend. But there are a few that I suggest in any case

1.Cleaning for lawn region: Nothing like taking your garden hose and cleansing the gardening place with a jet spray of water. It makes the area appearance neat, it prevents deposits of leached potting combo from pots.

2.Potting and Repotting: As a gardener, you want to have skilled the temptation of potting your nursery-bought flora asap. The weekend is the top notch time to pot your nursery-bought flowers because you have got have been given extra time to put together, buy materials and interact in this onerous undertaking.

3.Adding components: You need to add herbal additives at regular durations and the great days to do are weekends. So whenever you are due for together with your components, do them on weekends.

4.Do-it-yourself fertilizers and pesticides: Best time to make your foliar sprays so that you can save them and use, recipes you may find out in Chapter 6. Secrets Of Feeding Your Plants and Chapter 7. Organic Pest Control

Weekly 3: Weekly feeding if any

In Chapter 6. Secrets Of Feeding Your Plants I clearly have shared many do-it-your-self natural fertilizers and in Chapter 7. Organic Pest Control many organic pesticides as properly. If you may make the ones organic fertilizers, then you can set a weekly schedule for its software:

1.Cow dung + Cow urine foliar spray

2.Moringa foliar spray

three.NEEM & WD PESTICIDE/NEEM OIL

4.Masala pesticide

Give as a minimum three-4 days hole among for feeding & pesticide foliar spray to superb that its uptake through way of flowers. For example

Saturdays: Neem Oil Spray or anybody aforementioned

Tuesdays: Moringa foliar spray or all and sundry aforementioned

Bi-Weekly regular

Once in each 2 weeks or every 15 days. This is an non-compulsory time table to set so experience free to pass it. However, in case you intend to use any of the following, then you definately need to do them right here:

1.Micronutrients foliar spray

2.Liquid manure the usage of kitchen waste

3.Moringa foliar spray

I absolutely have shared the recipes in preceding chapters.

This is likewise an outstanding time to shift some of your weekly obligations for example Moringa foliar spray (if you made one)

Monthly ordinary

There are few activities which can be properly ideal to do on monthly basis, right right here's what I do:

Monthly 1: seaweed extract spray

Seaweed is taken into consideration an outstanding all-herbal boom booster. When farming it's far advocated on/earlier than particular levels like pre-flower, fruiting, and so on. To boost up growth. There is not any harm in following this pattern too to your terrace garden. However spraying it on month-to-month foundation might also have an brought benefit for your flowers and this is micronutrients, seaweed extract is an remarkable supply of micronutrients and therefore I will act as a monthly multi-food plan dose on your flowers. Few topics to be conscious at the same time as spraying it:

1.Even if the bottle of seaweed extract says 2-three ml in a single-liter water, use 1ml with 1-liter water. I actually have seen people unfavourable their vegetation with more quantities.

2.Follow the recommendations of foliar spray as referred to in 7 commandments of foliar feed.

three.If it is too heat a day, then surely bypass it the following day.

Quarterly normal

Container-grown vegetation want normal feeding and sluggish-launch fertilizers paintings wonders for them. In Secret four: rejuvenating potting mix I clearly have described potting combo rejuvenation and its recipe. You need to upload herbal components on your plants on the begin of the subsequent months:

1.June

2.September

3.Dec (in warmer/tropical areas)

4.Feb-March

Specific feeding necessities for plants and their dormancy is past the scope of this e-book. Therefore hold a have a look at of this:

Don't feed dormant plant life: While you may want to do it for all of your subject-grown plant life, maintain in mind that a few flora do not want to be fed of their dormancy season it in reality is maximum probably wintry climate/snow/frost season.

Winter in no manner truly comes: If your location does not revel in snow or frost you may experience pretty some fruits/veggies/vegetation that can be grown in wintry weather season and consequently you can upload additives to those flowers.

Growing season everyday

The time of 3 hundred and sixty 5 days at the same time as a specific plant (plants, cuttings, seedlings) suggests rapid boom, is known as growing season in clean terms. A growing season provides an most first-rate surroundings for the plant to increase that's why some gardening sports sports are done in particular in the route of the growing season.

Note: I am not speakme about seed sowing season right right here that is beyond the scope of this e-book. You can also additionally seek the net to outcomes find endorsed sowing months for a selected sort of veggies/cease result/plants.

For maximum gardeners and of path for max plants, there are growing seasons in India:

1.Spring Season- Feb-March (early), Early April in a few regions.

2.Rainy Season - May give up to June (till July in a few regions).

Growing season 1: Pruning

When a plant's dormancy duration ends it suggests signs and symptoms of new boom. That is why whilst you prune a plant it suggests new boom in phrases of latest leaves and branches. For a majority of flora, spring is the wonderful time to prune. The tremendous time to prune within the spring season is from the first week of Feb to the third week.

Why prune?

1.Pruning makes a plant dense through manner of selling new branches. More branches mean more culmination and plants. This is especially actual when you have fruit wooden on your terrace garden.

2.Pruning is an opportunity to get rid of antique/useless/damaged or malformed (like one over each exclusive) branches.

three.Some vegetation can simplest fruit/flower over new branches for instance apple ber (Indian jujube)

When to skip pruning?

1.If your plant is simply too more youthful like to procure it from the nursery or it's miles however setting up in its box. Otherwise, it is advocated to prune greater youthful plant life for promoting growth.

2.If your plant is a sluggish grower.

three.If your plant is not healthy.

Hard prune or mild prune?

When you difficult prune a plant you do competitive pruning and go away the plant in a reduced duration with all branches (smaller in addition to larger ones) lessen to decreased duration. Usually performed on well-established and matured vegetation.

A hard pruned plant

A hard pruned Indian jujube tree in the month of May with mulching for moisture retention

New increase inside the beginning of monsoon after the pruning and mulch eliminated

When you soft prune a plant you cut the smaller branches, looking for to preserve the shape of the plant. Soft pruning might also moreover contain defoliation (elimination of leaves) to sell new boom but it isn't always continuously essential. Depending on plants easy pruning can be completed almost any time of the 12 months, as an example, you can soft prune a bougainvillea plant after every flowering cycle to sell extra flowering branches.

Growing season 2: Repotting

When you want to shift your area plant from one container to a few one-of-a-kind typically a larger one (but not constantly), its called repotting

Why repotting?

1.Roots are popping out of the drainage hole at the bottom of the box.

2.Roots are growing so thick in the container, they may be de-shaping the world.

3.Signs of stunted increase or slower growth than ordinary.

four.The famous shape has emerge as massive and heavy such that it could imbalance the field.

five.Leaves are becoming wilted out rapid after watering and need greater not unusual watering.

6.Salt deposits on flowers or the sphere.

7.Soil is shrinking inside the container.

eight.To rejuvenated or update potting mixture.

Repot only in the developing season?

No, and sure.

NO, because of the fact most of the motives noted above may additionally require at once movement to maintain your plant. For instance, your plant is overgrown or root sure or now not keeping water. Please repot.

In this kind of case I endorsed taking the whole ball out and in reality transferring it to a larger box with smooth potting combination. Lesser or no damage to plant and higher recovery price.

YES due to the truth repotting is usually demanding to plant and repotting a plant at the begin of the growing season guarantees steady restoration of the plant. Also, a few plants grow better at the same time as you re-pot them periodically like every year or once in 18-24 months.

How to repot your plant?

In my experience we're capable of divide the repotting technique into two types as follows:

1.Extract and Repot

2.Extract, Reduce & Repot

Since every the kinds have an extraction, extraction can be defined as cautious elimination of plant from its field.

My mulberry plant needed repotting because of the truth it is extraordinary root positive

How to extract?

1.To extract a plant from the field, stop watering the plant a day or in advance than, this could introduce moderate compaction inside the mix and permit the combination to transport a long way from the field's internal wall.

2.Then lightly pat the sector from all sides to further loosen the complete ball.

3.If you have got accompanied my recommendation for larger drainage holes, you can even gently use your fingers to push the ball from the lowest.

4.Use gravity that will help you if crucial thru turning the plant the other manner up, but be aware that soil may additionally fall out.

5.If you have got a better platform/desk/stool then location the field mendacity down and lightly pull the plant keeping it from the stem, in case you sense that stem isn't tightly certain, forestall and attempt to provide moderate pushes to the box, making it barely tilted from backside.

6.As a final step, gently do away with the plastic internet/material/ or if you have used damaged clay pot portions from the ball. It is ok if tiny roots harm a touch but now not too much.

7.A free textured and well-tired combination will fall on its very personal without horrible the roots, in any other case, lightly brush them to loosen a bit only within the occasion that they appear tangled in any other case leave them as it's far. This will provide some air to the roots in the new subject and provide area for the mixture.

A well worn-out potting mixture is simple to shred

Extract and Repot

When you extract a plant and repot really the way it's miles in a brand new huge discipline with the fresh combo to deal with the space after setting the entire root ball as it's miles. Here's the way you do it:

1.Prepare your new box preferably huge with a few potting mixture and vicinity the entire ball inner such that even as you upload extra soil, it should cover the ground to the stem as a minimum an inch.

2.Gently poke arms or some stick anywhere within the newly introduced combo to easy air wallet and add more mixture. Keep tapping the arena from the edges too.

3.Don't fill the field until its neck. You want some room for water to soak slowly into the field.

4.Don't % the potting mix too tight by way of pressing it collectively at the side of your palms, do it lightly to offer guide allow the ultimate paintings be performed with the aid of manner of watering for you.

five.Finally, water the plant generously and that's it.

After repotting is completed

Extract, Reduce and Repot

The extraction and repotting strategies are without a doubt the same with one huge difference. Reduction. In this technique you can maximum possibly do all or some of the subsequent:

1.Shedding as lots vintage potting blend as feasible without hurting the foundation ball connected to the stem.

2.Trimming fibrous roots which is probably overgrown/tangled/broken/rotted.

An antique jasmine plant potted in clay like lawn soil, I must use a observed to trim the roots and decrease it.

3.Trimming the entire ball from the edges and backside sacrificing a number of roots all throughout.

four.Once you do that, then spray the roots with a few fungicide preferably natural like crushed and boiled garlic diluted with water in a ratio of one:10 or sprinkle trichodderma powder (to be had at online stores).

5.Prepare a modern-day field with the easy potting mix, place the decreased ball, and cover it with greater potting mix gently pressing from all sides, poking the combination with hands or a timber live with put off air wallet.

6.Make fantastic the ball is properly stabilized and now not tilted.

7.Water the plant.

8.Allow the combination to dry amongst subsequent watering. For some flora, you could want to sit up for very prolonged durations in advance than you water after repotting just so the roots are nicely settled for instance Jade plant, Adenium, or similar succulents or flowers unique to such remedy.

Protection from extra heat

Summers in sure elements of India may be definitely heat. So warm that even virtually grown plant life/flowers may also dry to lack of life inside the absence of proper hydration. No marvel why poly-residence farming is growing every day because it lets in controlling a number of situations like temperature, light, humidity required for particular vegetation.

Container-grown flowers need protection from warmness for numerous reasons, beneath are the most important ones as consistent with my revel in.

Chapter 10: Things You Need To Start

A terrace garden is surely now not a luxurious in case you understand properly approximately the listing of necessities you need to setup a lovely inexperienced space to your terrace. Most of the necessities that you want to setup a rooftop garden are with out troubles available in all nurseries and moreover many huge supermarkets sell them. This monetary break is all about the requirements you need to setup a lovely terrace lawn irrespective of the scale and the shape of your terrace.

Important Note: Even earlier than intending with terrace garden necessities, the following attitude is essential and specifically recommended so that you will sense encouraged until your lawn receives proper into a proper form. Just keep a be aware of the following matters:

1.Be ready to spend at least one dedicated hour each day for your terrace lawn. Once the setup begins offevolved operating properly you can spend great little even as jogging there.

2.Get organized to get your hands dirty via going for walks in soil, water, manure and so on.

3.Make way for a beneficiant mind and be organized to welcome a few lovely and not-so-adorable insects, insects and so forth.

4.Do not reflect onconsideration on growing too many things at a time. Make a list of what is going even as after which take movement.

5.Be prepared for few disasters on the primary month. Ensure that you don't surrender but research from your errors.

6.Do enough net research and preserve studying so you hold close extra moderen tips and increase better hacks.

7.One rotten plant can reason many because of this be ready to dispose a rotten plant even though it is your selected.

Now which you are mentally organized, allow's get into motion! Here is a tick list of things you want to setup a terrace garden:

1.Waterproofing and power checking – The first trouble you have to do to start your terrace garden is to water-evidence your terrace. To begin with, you could buy a thick tarpaulin sheet, this is unfastened from tampering or holes and cowl the floor ground of your terrace. This method works simplest if you are inclined to have pots for your terrace. If you want to do floor gardening then name for professional help to water-proof your terrace. After sealing the terrace place, take a look at in case your foundation is robust enough to preserve the extra weight of your terrace garden. Mostly the solution can be a "certain".

2.Buying soil – You want to find out a correct supply from in which you can buy soil. Most of the nurseries sell nicely-nourished soil that could be a mixture of soil, vermicompost, and compost coir. If you discover rich crimson soil near your home then you could don't forget using it too by way of method of mixing it with sufficient vermicompost and compost coir. The remarkable possibility for purchasing soil is the soil from your private floor if you have a ground garden in your property. Your out of doors soil is also a terrific preference if it is not polluted with chemicals like detergent, oil and so forth.

three.Water – Since it is terrace, water will now not be a huge challenge. Just get a hose with sufficient length to supply water to the flowers for your rooftop lawn. Alternatively, for smaller plant life you could consider a small or medium period watering kettle. For higher water retention in smaller plant life you could remember artificial sand that comes inside the form of silicon balls. This retains water for an prolonged duration (that is only optionally to be had for placing pots).

4.Pesticides – Garden pests will constantly make manner into your terrace garden regardless of the security you provide for your lawn. It is suggested that you hold primary insecticides in hand to fight the invasion of pesticides. Say a strict 'no' to artificial pesticides that is absolutely toward organic farming at home. You can put together few herbal pesticides via following the commands stated beneath:

Mix baking soda, lemon oil or orange oil, little little bit of dish cleansing cleaning soap, cooking oil, garlic juice, chili powder and water in a sprig bottle and use it as a not unusual pesticide.

Mix cooking oil, baking soda and one quart water in a twig bottle and spray it on flora to keep away from fungal infections.

Place a small bowl complete of a hundred fifty ml beer in it. This draws snails and snugs invading your garden and the snails receives into a drunken oblivion simply so it is straightforward to move them far from your garden.

5.Planters and boxes – Planters and boxes are the maximum crucial phase of your lawn. Based at the layout and the style of rooftop gardening you follows the boxes and planters may be chosen for your garden. For a rooftop garden, nearly some issue you could determine as a planter like a bottle; UV dealt with developing luggage, unused bins, helmet, vegetable crate or drinks crate, sure types of waterproof wood containers and so on. And something that is able to retaining water and unique developing substances in them. Apart from DIY planters and pots there are precise planters and pots available inside the nurseries which might be made from clay, dirt, cement, terracotta, cement and lots of others. Except for the cement planters choose out some thing which you need on your terrace lawn.

6.Garden covers – There are unique types of garden covers manufactured from nylon and mesh are available to guard your smaller fruit bearing and vegetable bearing flowers from birds. Buy most effective few and keep them organized to shield your smaller plants and seeds.

7.Veg trug – To grow vegetable flora, if you have the finances then you can choose a veg trug too. These veg trugs can preserve water and they will be far from the concrete surface and they will be the great to broaden greens and shrubs.

eight.Pot reservoirs – If you revel in your terrace receives maximum sunlight hours and your flowers need extra water retention functionality then reflect onconsideration on few pot reservoirs in your terrace.

nine.Pot hangers/holders – How about a terrace garden without hanging pots with colorful flowers? It sounds unpleasant isn't it? Then get a few lovely pot hangers, in particular for creepers and flower bearing terrace plants.

10. Seeds – Stake few essential seeds like tomatoes, girls finger, carrots, cauliflower, oregano and plenty of others. That you may

with out issues increase for your terrace as a number one time interest gardener.

Now that you have stored the entirety you need to begin your terrace lawn so in the next bankruptcy permit's get began out out with rooftop gardening methods.

Chapter 11: Steps Involved in Setting up Terrace Garden

Step 1: Stocking the requirements

The first step concerned in putting in a terrace garden is to inventory at domestic all essentials you need to setup a lovable rooftop lawn. Now which you have the whole lot listed within the first bankruptcy for a garden so observe the following vital steps to create your adorable rooftop lawn.

Step 2: Install drainage and wind boundaries

Rooftop lawn is placed in a windy region, immoderate wind typically generally have a tendency to make the delicate stems of your extra youthful flora damage. Installing a wind barrier is absolutely important at this degree. If you've got the budget then setup a wind barrier similar to those tested beneath.

The wooden wind barrier can be little highly-priced that is why a secondary opportunity of looking for twig fencing is suggested as a wind barrier in a rooftop garden.

After the wind barrier, search for a proper water drainage tool on your terrace. The greater water from the garden may be result in rain water harvesting pipe just so the water reaches the ground straight away enriching your ground water system.

Step 3: Prepare the layout

Soon after the wind barrier is set up, it is time as a manner to put together the layout to your garden. Think of a layout with better drainage device, clean and to be had to water the flora, the format ought to be accessible enough for you to acquire each plant. Ideally look for a shady layout in order that weaker flora can be saved in a shady region (i.E. Close to the terrace partitions). Stronger and warmth resistant flora may be stored in a sunny place of your terrace. The notable format for a terrace garden may be a square format with herbs in a unmarried line, vegetation within the facets and vegetables in a line parallel to the herbs. A center patch is also neat if you need to setup a vegetable trug in your terrace lawn.

Step 4: Set up the boxes

You can also additionally even use the antique bathing bath of children made from UV treated polythene as a garden patch for herbs that calls for an awful lot plenty less root area. Get a waterproof polythene sheet and cover the concrete floor of your terrace just so excess water from your dust pots and unique pots don't stagnate or penetrate inside the concrete roof of your private home. Initially, hold packing containers with rails just so it is straightforward so as to flow into your smaller plant life to shady or a high-quality deal tons much less windy location of the terrace frequently.

Step 5: Prepare the soil

A plant can't stay to inform the tale most effective with soil and water; it dreams sufficient vitamins to live to tell the tale and to live nourished. This is why potting combination or potting media is suggested. To get the proper potting blend follows the steering given underneath:

• Take 5 kg of soil (ideally red soil or your lawn soil that is chemicals unfastened)

• Add sheep, goat or cow dung to the soil. For five kg in truth nicely well worth soil, the dung combination want to be for approximately 500 gms

• Get dry cow dung cake, burn it and upload the ash to the soil inside the container.

• Add neem oil to the soil on the manner to prevent it from insects and germs.

• No add creation soil weighing round 10 kgs to the mixture.

• Now you may add coconut fiber to the mixture to make it stronger. Coconut fiber has no vitamins but they may maintain water for an extended time span consequently your plant life want them.

• Finally water the aggregate, add vermicompost to the combination and hold it apart for approximately 3 days to settle via continuously watering the combination.

Let this aggregate be given about three days with the beneficial resource of watering them regularly in order that your nutrition rich soil is prepared for planting. While getting ready the soil within the packing containers, it is important which will make certain the drainage device of the field. The drainage want to now not allow go off all the nutrients from the soil combination through being large. Also at the same time, the holes to your discipline shouldn't be very small so that water stays within the container. In case, if the hole on your location is in reality too huge then cover it with coconut fiber, cord mesh or using a cotton fabric.

Step 6: Plant your seeds -Planting

After three days of settling the soil, it is time as a way to plant your seeds. Planting can be achieved in levels. The first section is planting your seeds in a seeding tray and the second one phase is ready shifting the saplings to the pot efficiently. Invest in a tremendous seeding tray that is entire of the enriched soil with all the combos inner it in a advocated quantity after which begin planting your seeds inside the tray. Ensure your seeds are properly planted; normally a area of approximately inches needs to take shipping of for the seeds to grow and attain the pinnacle-most layer of the sand. Let the small plant or the sapling make bigger a touch so you can carefully transit the equal to a eternal pot or a subject. Sometimes look forward to five to 6 leaves to spring out so that you recognize that plant is properly nourished and is coming out nicely so transit can be easier.

As speedy due to the fact the seeding is over, you can look ahead to about one week for the smaller stems to expose off on the seeding tray, now allow the plant expand for about three or four inches in height and then transit it to a separate pot. While transferring the plant to a pot from a seeding tray, ensure you don't reveal the foundation to direct daylight hours because of the truth warmth can immediately permit the roots dry simply so the roots die. Take the seeding tray to a shady place of the house after which fast transfer the plant into the field.

An expert way to transit plant life to a area is to go away some sand at the side of the premise in order that even you can't see the basis after which fast area the sand with plant into the box and then too seal it with greater sand in the new container.

Step 7: Pest control

Now that your terrace lawn is prepared and is coming out exceptional, as an owner of this lovable lawn you want to deliver lower back pests out of your lawn with proper pest manage capabilities. Pests would like to stay on your lawn even if you dislike them..

Chapter 12: 6 Best Terrace Garden Methods

Setting up a regular garden is exceptional approximately identifying the panorama of the garden. You virtually need to determine the decorative like fountain, planter affiliation, plants association and so on. In a everyday garden. When it involves terrace garden, figuring out the garden method is proper to keep with unique steps of rooftop gardening. In our preceding chapters we've got mentioned about the requirements you need to setup a cute rooftop lawn and we additionally referred to approximately a manner to start with gardening.

In this bankruptcy, we may also want to communicate in detail approximately the 6 particular techniques concerned in terrace gardening. While your plants are in a budding stage itself, if you make a decision which terrace gardening approach you're going to study, it's miles going to be smooth to a good way to preserve your rooftop lawn with utmost knowledge. This financial disaster will provide an cause of you approximately the brilliant terrace lawn strategies which might be suitable for growing vegetables, cease quit end result, herbs or perhaps flowering plant life.

Method 1: Pot Garden on Terrace

The maximum common and the very best manner to setup your personal terrace garden is to have a vital garden setup with pots. Setting up terrace gardens with pots is straightforward, available, a good deal less time ingesting and that they purpose plenty a good deal less damage to the roof of your concrete building when you have nicely completed water proofing in your ceiling.

There are stunning, modern, in your price range, robust and one-of-a-type sizes of pots to be had inside the market. There are clay, terracotta, concrete, marble, plastic, melamine, or even China clay pots are available inside the market. Excluding the marble pots, the entirety else is a good deal much less expensive and they weigh lots much less for a rooftop lawn. Almost all commonplace veggies like chilies, peppers, onions, tomatoes, gourds, beans, beetroots, carrots and masses of others. May be grown in pots on a rooftop lawn. The biggest gain of a pot lawn on terrace is that they may be smooth to move and rearrange to exchange the advent of the lawn every time.

Method 2: Raised mattress terrace gardens

The first-class gardens for terrace especially if you want to broaden greens are raised mattress rooftop gardens. Raised beds are perfect for developing greens because of the reality they're spacious, they've got even soil, it is easy to develop a combination of plant life in separate or equal boxes that helps you to test developing hybrid vegetables.

Raised bed gardens are smooth to maintain due to the reality the soil does not get too company so that you need to loosen it. The setup for raised bed garden is likewise low-cost and without problems available at maximum of the nurseries and at the community hardware stores. Get the assist of a close-by wood worker to setup raised beds in your terrace to get the landscape of your rooftop garden right after which setup a lovely raised mattress garden for your terrace.

Method three: Integrated patch on terrace

Integrated vegetable patch rooftop garden approach is quality for large and spacious terraces. The included vegetable patch is a exquisite way to grow veggies like broccoli, cauliflower, cabbage, and herbs. These patches will make your terrace appear to be a really perfect floor lawn.

Integrated vegetable patches are quality for developing seasonal veggies and end result in your terrace and they may be mild weight too while in evaluation to different rooftop lawn sorts. All you want to make certain to have a a hit rooftop garden with covered vegetable patch is to make certain the air waft; wind control and drainage of the lawn are ideal.

Method four: Terrace gardens with planters

Terrace gardens created the use of colorful planters look incredibly attractive and extraordinary. These planter bins are apt for flower lawn to your rooftop that may be great used for aroma therapy. Colorful planters are available in unique shades, shapes, sizes and additionally they'll be custom designed as well with railing to regulate your terrace garden panorama regularly.

are the right desire in case you need to create a terrace garden that looks rustic, innovative with extra software place. With planter containers setup to your rooftop, you could pass them effortlessly the usage of the railing to create region in your rooftop barbeque party, a cushty studying session on a breezy day or maybe to set up your dinner for a domestic date.

While installing the lawn with planter containers, make sure that excellent the notable of compost soil and specific manure goes indoors them to make a genuinely high-quality aggregate of nourished soil your plants want. Planter bins can be hung as properly to make you terrace lawn appearance greater specific. Most not unusual plants that can be grown in a terrace garden made with planter packing containers are flowering flowers, spices, herbs, and also few leafy vegetable flowers like garlic, tomato, cucumber, chilies, bush beans and plenty of others.

Method five: Vertical terrace gardens

The optimized vertical region in your terrace may be implemented for a green lawn at your rooftop. These vertical terrace gardens are apt for growing flowers and herbs. In a vertical terrace lawn an unused wall can be used for assist and a cord mesh may be used for vertical help of the vines. The terrific vegetation aside from flowering plant life to extend in a vertical terrace lawn consist of shrubs, vines like gourds, beans, squashes, few types of peppers, chilies and even particular vegetable sorts that goals the useful aid of the wall to develop.

Vertical terrace gardens may be applied to store more region by using letting the flowers develop across the wall for that reason they're best for compact terrace gardens.

Method 6: Rooftop gardens with repurpose sandboxes

A decorative repurpose sandbox on your terrace isn't always super for children to have fun with sand but they may be implemented to create a scenic green terrace too. The repurpose sandboxes make a great mini terrace lawn thru way of presenting excellent area for the plant life to increase and nourish properly. Imagine a cute repurpose sandbox full of aromatic herbs to your terrace in a shady region with a chair nearby as a manner to experience your chosen e-book and a cup of coffee? This seen is bliss isn't it? Then right now purchase a repurpose sandbox and start prepping for a adorable terrace garden.

Repurpose sandboxes are a first time gardener's amazing buddy due to the fact they help you keep more region and test with a substantial form of vegetable plant life like radish, lettuce, garlic, pepper, ginger, oregano, basil and extra.

You can remember cutting-edge techniques to installation a location savvy terrace gardens thru using your antique bathtubs, painted tires of the cars, using inverted umbrellas as putting pots, vintage water kettles as planters and extra. These small add ons for your terrace will upload colorful sunglasses in your terrace lawn and purpose them to look appealing.

Chapter 13: Perfect Plants to Grow at the Terrace

Now that you apprehend approximately the requirements to setup a terrace garden, the manner to keep it, a manner to landscape your terrace lawn and greater. It is time that lets in you to moreover find out about the plants you may develop in your terrace garden so you can invest best in result-yielding seeds and saplings. As a newbie, this guide will can help you pick the right flowers you may increase for your terrace just so you may be sadness unfastened. One excellent tip to preserve in mind whilst choosing your flowers for terrace lawn is to put money into larger planters or spacious pots, patches or sandboxes in order that flora increase freely and furthermore large planters have a tendency to hold moisture for an prolonged time span.

Vegetables for terrace gardens

Vegetable terrace gardens need greater care. They need adequate water, right drainage device, slightly acidic soil, and specially, the soil need to we extremely fertile. Little or more daylight hours is ok for a vegetable that can be raised the usage of raised terrace mattress, repurpose sandboxes or maybe using smaller planter.

As a number one timer you may strive growing smooth-to-boom veggies on your terrace lawn like chilies, tomatoes like cherry tomatoes, cucumbers, or maybe okras. Once you flavor fulfillment in developing the salad flowers then you may flow into right away to vegetables collectively with eggplants, potatoes, peppers, radishes and so forth.

Succulents and perennials for terrace gardens

If you are first time hobby terrace garden owner then start thru growing succulents and perennials in your terrace. These succulents and perennials are beauties amidst terrace plants due to the fact they in no way will be inclined to dry out quicker because of excessive warmth, daylight and wind exposure.

Grow succulents and perennials in your terrace garden first like lilies, marigolds, ferns, cacti types and lots of others. You may even don't forget grasses, sage, and composite family plant life like chamomile, asters, daisies and masses of others. These flora require low protection and they are able to develop nicely in dry situations too. However begin with succulents and perennials after which flow into to vegetables and herbs.

Flowers for terrace gardens

If you preference to develop plant life on your terrace lawn you then need to consider growing flowers which may be annuals. Grow plants along with marigold, pansies, begonias, petunias, lilies, daisies, hibiscus, hyacinths, and additional in your terrace lawn. You can don't forget seasonal plants like jasmine, roses, colourful bulbs that embody primroses and plenty of others. In your rooftop terrace too.

Fruits, Vines and Dwarfs for terrace gardens

Most of the dwarf fruit bearing vegetation can be grown in a terrace lawn but the ones stop end result and flowers require maximum care and preservation. Dwarf fruit wood like apple, pomegranate, guavas, figs and so forth. May be grown to your terrace. However lemon tree may be a lovely addition to any terrace garden. Based on the humidity stages you can broaden grape vines, olive timber, mangoes additionally on your terrace. Even short bonsai wood make a lovable addition to a planter based totally totally terrace garden that receives sufficient shade and mild.

Plants for herb terrace gardens

Herb terrace lawn can be aromatic; they're a awesome desire for a primary time terrace gardener. You can begin your herb terrace lawn with a pot or a container with basil and whilst you find it growing nicely then you could waft straight away to a number of herbs like rosemary, thieves, chive, parsley, mint, cilantro, lemon grass and greater.

Chapter 14: 7 Secrets for developing pleasant terrace gardens

Is your intention is to convert your terrace proper right into a lush green lawn complete of colorful end result, plant life and veggies and fragrant herbs? Then the time tested secrets and techniques with the beneficial aid of experts listed on this chapter to maintain a lovely terrace lawn will genuinely prove to be useful to you. The maximum vital key to elevate a adorable terrace lawn is to realize the right blend of soil, sunlight hours, water and information which plant to develop whilst. The following are the ten splendid secrets of specialists who raised stunning terrace garden in their homes.

Get the right region – Not every terrace or rooftop is meant for terrace gardening, you have to first get your place right in case you need to elevate a wholesome terrace lawn. The vicinity for terrace garden need to look easy, clean, they location should be satisfactory for growing all varieties of flora which includes smaller plants that might spoil effortlessly if the place is so windy.

Sunlight - Sunlight is each different important detail for a fitness terrace garden sooner or later chooses a place that gets enough sunlight hours in your flowers to live on. Not an excessive amount of of daytime but surely enough daytime for approximately to 3 hours is extra than enough to your flora to stay to tell the story. During summer season, try to avoid setting herbs planters in direct daylight specially inside the afternoon solar because robust rays from sun penetrate deep inside the soil in patches, planters and containers. When this takes place the moisture of your soil can be taken away as a result the roots generally tend to dry out faster. Understand if the roots dry then the possibilities of survival for the plant is too much much less.

Right mix of soil — Irrespective of the dimensions, form, and method of gardening you're planning to comply with and further you need to get the proper combination of soil in your lawn. Ensure the soil you are the use of is enriched the use of enough compost mix, vermicompost, natural fertilizer, coconut fiber and so on. AS defined inside the early chapters you have to prepare your soil right and vitamins wealthy to get the soil proper for any form of plant.

four.Restore soil after rains – The mistake maximum of the primary time gardeners do is best in the path of the monsoon. After a heavy rain, human beings will be inclined to head away their lawn as it's miles with out yet again enriching the soil in their lawn, that could be a key motive behind why flora pass below nourished and don't supply predicted yield. It is constantly critical that permits you to make sure the soil in your lawn is properly nourished after the monsoon or towards the give up of the monsoon.